Springer Theses

Recognizing Outstanding Ph.D. Research

Aims and Scope

The series "Springer Theses" brings together a selection of the very best Ph.D. theses from around the world and across the physical sciences. Nominated and endorsed by two recognized specialists, each published volume has been selected for its scientific excellence and the high impact of its contents for the pertinent field of research. For greater accessibility to non-specialists, the published versions include an extended introduction, as well as a foreword by the student's supervisor explaining the special relevance of the work for the field. As a whole, the series will provide a valuable resource both for newcomers to the research fields described, and for other scientists seeking detailed background information on special questions. Finally, it provides an accredited documentation of the valuable contributions made by today's younger generation of scientists.

Theses may be nominated for publication in this series by heads of department at internationally leading universities or institutes and should fulfill all of the following criteria

- They must be written in good English.
- The topic should fall within the confines of Chemistry, Physics, Earth Sciences, Engineering and related interdisciplinary fields such as Materials, Nanoscience, Chemical Engineering, Complex Systems and Biophysics.
- The work reported in the thesis must represent a significant scientific advance.
- If the thesis includes previously published material, permission to reproduce this must be gained from the respective copyright holder (a maximum 30% of the thesis should be a verbatim reproduction from the author's previous publications).
- They must have been examined and passed during the 12 months prior to nomination.
- Each thesis should include a foreword by the supervisor outlining the significance of its content.
- The theses should have a clearly defined structure including an introduction accessible to new PhD students and scientists not expert in the relevant field.

Indexed by zbMATH.

More information about this series at https://link.springer.com/bookseries/8790

Yoko Oya

A Few Tens au Scale Physical and Chemical Structures Around Young Low-Mass Protostars

Doctoral Thesis accepted by
The University of Tokyo, Tokyo, Japan

Springer

Author
Dr. Yoko Oya
Department of Physics
The University of Tokyo
Tokyo, Japan

Supervisor
Prof. Satoshi Yamamoto
Department of Physics
The University of Tokyo
Tokyo, Japan

ISSN 2190-5053 ISSN 2190-5061 (electronic)
Springer Theses
ISBN 978-981-19-1707-3 ISBN 978-981-19-1708-0 (eBook)
https://doi.org/10.1007/978-981-19-1708-0

This Springer imprint is published by the registered company Springer Nature Singapore Pte Ltd.
The registered company address is: 152 Beach Road, #21-01/04 Gateway East, Singapore 189721, Singapore

Supervisor's Foreword

In the late 1960s, molecules such as ammonia, water, and formaldehyde were successively discovered in interstellar clouds through radioastronomical observations. It was surprising that such polyatomic molecules familiar to us are abundantly present in a harsh condition of an interstellar space. This discovery excited astronomers to dream about its relation to the origin of life in the universe. Since then, more than 200 interstellar molecules including various organic molecules have continuously been detected, thanks to the advances in radioastronomy. Indeed, the universe is much richer in molecules than ever thought. Among them, discovery of long carbon chain molecules such as cyanopolyynes was very surprising, and the research reproducing them in the laboratory led to the opening of a splendorous world of fullerenes and carbon nanotubes.

However, the astronomers' dream of exploring the origin of the Solar system, as well as the origin of life, did not proceed so rapidly, because it was difficult in those days to trace how molecules formed in interstellar space are brought into planetary systems. A typical size of the disk structure where a planetary system is being born is 100 astronomical units (hundred times the distance between the Sum and the Earth), and the disk structure is deeply embedded in a thick envelope of gas in its infant stage. This size corresponds to the apparent size of the plane of the sky of less than 0.7 s of arc for nearby star-forming regions (450 light years). Hence, high spatial resolution and high sensitivity are required for molecular line observations to characterize its physical and chemical structures. Nevertheless, there were a few progresses during 2000s. Fairly large organic molecules such as methyl formate ($HCOOCH_3$) and ethyl cyanide (C_2H_5CN) were detected in the vicinity of the protostar with sensitive radio observations. This result means that substantial amounts of organic molecules (even larger molecules) are already formed in interstellar clouds before the formation of the planetary system. Hence, the above discovery attracted broad attention of astronomers and planetary scientists in relation to pre-solar organic materials found in meteorites. Moreover, chemical composition of protostellar cores was found to be different from source to source, which raised another important issue to be studied, that is how such chemical diversity is inherited in a planetary system.

ALMA (Atacama Large Millimeter/submillimeter Array), which started operation in 2011, was really a game changer. ALMA realized an unprecedented angular resolution in the millimeter- and submillimeter-wavelength bands with enough high sensitivity to detect various molecular lines. With ALMA, the central part of protostellar sources where the disk structure is being formed is spatially resolved, and a chemical structure of the disk-forming region can be studied in relation to the physical processes of disk formation. Such studies are now being expanded to many sources as a new trend in star-formation study and astrochemistry. Moreover, satellite missions exploring comets and asteroids were successfully conducted, and the chemical composition of pre-solar materials is now being investigated. Thus, the astronomers' dream is becoming true little by little through these advances.

This Ph.D. thesis presents a frontier work in this direction. Dr. Oya studied the physical and chemical structures of the five young Solar-type protostars with ALMA. The novel aspects of this study are threefold. First, Dr. Oya focused not only on a physical structure, but also on a chemical structure and clarified their relationship. She successfully revealed that a specific part of the physical structure in the disk-forming region can be highlighted by specific molecules. This means that molecular lines can be used as a 'molecular marker' to study physical structures. Moreover, the chemical diversity found in protostellar cores is indeed inherited in the disk-forming region. Second, Dr. Oya combined the analyses of a disk/envelope structure and an outflow. These two components have traditionally been studied by different groups specialized to each component. However, they are mutually related to each other through the angular momentum of gas particularly in the vicinity of the protostar. She effectively used the information on the outflow for the disk/envelope analysis and vice versa. Third, Dr. Oya revealed that the observed physical structures can be interpreted by simple physical models for a disk, an envelope, and an outflow. These models reasonably represent the basic structure of these components and allow her to disentangle the complex chemical feature of disk-forming regions.

In short, all these results opened a new avenue in star-formation study and astrochemistry by revealing the physical and chemical evolution of disk-forming regions in the early phase. It is amazing that such a comprehensive work was completed in four and a half years of her Ph.D. course. The success entirely owes to the effort and the talent of Dr. Oya. I can state without hesitation that this Ph.D. thesis indeed made a giant step toward the original dream, the origin of the Solar system and the origin of life in the universe. I am glad to know that this work is published as Springer Thesis. This publication is very timely, considering the rapid growth of this field. I believe that the content of this Ph.D. thesis will be useful for the star-formation and astrochemistry communities. I also hope that this Ph.D. thesis will also contribute to stimulating new science by researchers and students in broad areas of physics and chemistry.

Tokyo, Japan
February 2022

Satoshi Yamamoto

Preface

A thorough understanding of the star/planet formation and the associated chemical evolution is an important target for astrophysics and astrochemistry. Although radio and infrared observations have recently enhanced our understandings of the protostellar formation process, it is still controversial *when and how disk structures are formed around newly-born protostars*, and *how molecules of parent molecular clouds are delivered into the disks*. To answer these questions, physical and chemical analyses are conducted for disk-forming regions. This thesis describes the observational studies with Atacama Large Millimeter/submillimeter Array (ALMA) at a few 10 au scale resolution. Specifically, the following five young low-mass protostellar sources in the Class 0 and I stages are observed: L1527, IRAS 15398–3359, IRAS 16293–2422 Source A, Source B, and L483.

Rotational spectral lines of various molecular species are detected toward the above protostellar sources with ALMA. It is found that the kinematic structure of the flattened envelope at a few 100 au scale is well explained by a simple ballistic model. In this model (*infalling-rotating envelope model*), the gas cannot fall inward of a certain radius due to the energy and angular momentum conservation. This radius is called as the *centrifugal barrier*. In this thesis, the common occurrence of the infalling-rotating envelope and its centrifugal barrier is demonstrated for the five sources. A high-velocity component is also detected inside the centrifugal barrier. This component is attributed to the disk structure. Thus, the disk structures seem to have already been formed even at the earliest stage (Class 0) of the protostellar evolution.

The outflow components are also investigated in L1527, IRAS 15398–3359, and L483. Since the infalling-rotating envelope, the centrifugal barrier, the disk component, and the outflow components are thought to be related to one another through the angular momentum of the gas, their comprehensive understandings are essential to star-formation studies. The physical parameters (the protostellar mass, the radius of the centrifugal barrier, and the inclination angle) are indeed evaluated with the aid of the infalling-rotating envelope model, the Keplerian disk model, and the parabolic outflow model, and mutual relations of the components are discussed.

The chemical composition of the gas is analyzed in the above five sources. It is well known that the chemical composition of the protostellar envelopes shows significant diversity at a few 1000 au scale. In this thesis, it is found that the chemical diversity is indeed delivered from the envelope into the disk. Moreover, the drastic chemical changes across the centrifugal barrier, whose radius is typically a few 10 au, are also found. Three physical components of the disk-forming regions (the infalling-rotating envelope, the centrifugal barrier, and the disk component) are traced by particular molecules. These new findings allow us the chemical diagnostics of the disk-forming regions. The chemical changes across the centrifugal barrier are also of particular importance in astrochemistry of protoplanetary disks, because they are expected to determine the initial condition of the chemical evolution toward planet formation.

These results provide us with important clues to general understanding of the physical mechanism and the chemical diversity in the disk formation.

Tokyo, Japan

Yoko Oya

Parts of this thesis have been published in the following journal articles:

In reverse chronological order of publication.

1. Yoko Oya, Nami Sakai, Yoshimasa Watanabe, Ana López-Sepulcre, Cecilia Ceccarelli, Bertrand Lefloch, and Satoshi Yamamoto. "Sub-arcsecond Kinematic Structure of the Outflow in the Vicinity of the Protostar in L483". ApJ, 863:72, August 2018. arXiv: 1806.10298.
   DOI: https://doi.org/10.3847/1538-4357/aacf42
2. Yoko Oya, Kana Moriwaki, Shusuke Onishi, Nami Sakai, Ana López-Sepulcre, Cécile Favre, Yoshimasa Watanabe, Cecilia Ceccarelli, Bertrand Lefloch, and Satoshi Yamamoto. "Chemical and Physical Picture of IRAS 16293−2422 Source B at a Sub-arcsecond Scale Studied with ALMA". ApJ, 854:96, February 2018. arXiv: 1801.04174.
   DOI: https://doi.org/10.3847/1538-4357/aaa6c7
3. Yoko Oya, Nami Sakai, Yoshimasa Watanabe, Aya E. Higuchi, Tomoya Hirota, Ana López-Sepulcre, Takeshi Sakai, Yuri Aikawa, Cecilia Ceccarelli, Bertrand Lefloch, Emmanuel Caux, Charlotte Vastel, Claudine Kahane, and Satoshi Yamamoto. "L483: Warm Carbon-chain Chemistry Source Harboring Hot Corino Activity". ApJ, 837:174, March 2017. arXiv: 1703.03653.
   DOI: https://doi.org/10.3847/1538-4357/aa6300
4. Yoko Oya, Nami Sakai, Ana López-Sepulcre, Yoshimasa Watanabe, Cecilia Ceccarelli, Bertrand Lefloch, Cécile Favre, and Satoshi Yamamoto. "Infalling-Rotating Motion and Associated Chemical Change in the Envelope of IRAS 16293−2422 Source A Studied with ALMA". ApJ, 824:88, June 2016. arXiv: 1605.00340.
   DOI: https://doi.org/10.3847/0004-637X/824/2/88
5. Yoko Oya, Nami Sakai, Bertrand Lefloch, Ana López-Sepulcre, Yoshimasa Watanabe, Cecilia Ceccarelli, and Satoshi Yamamoto. "Geometric and Kinematic Structure of the Outflow/Envelope System of L1527 Revealed by Subarcsecond-resolution Observation of CS". ApJ, 812:59, October 2015. arXiv: 1604.08005.
   DOI: https://doi.org/10.1088/0004-637X/812/1/59
6. Yoko Oya, Nami Sakai, Takeshi Sakai, Yoshimasa Watanabe, Tomoya Hirota, Johan E. Lindberg, Suzanne E. Bisschop, Jes K. Jørgensen, Ewine F. van Dishoeck, and Satoshi Yamamoto. "A Substellar-mass Protostar and its Outflow of IRAS 15398−3359 Revealed by Subarcsecond-resolution Observations of H_2CO and CCH". ApJ, 795:152, November 2014. arXiv: 1410.5945.
   DOI: https://doi.org/10.1088/0004-637X/795/2/152

Acknowledgements

The author is grateful to Prof. Satoshi Yamamoto for his invaluable advice and encouragement. The author acknowledges Dr. Cecilia Ceccarelli for her kind collaboration. The author thanks Dr. Ewine van Dishoeck, Dr. Jes Jørgensen, Dr. Bertrand Lefloch, Dr. Ana López-Sepulcre, and Dr. Cécile Favre for their collaboration. The author is also grateful to Dr. Yuri Aikawa, Dr. Tomoyuki Hanawa, Dr. Tomoya Hirota, Dr. Masahiro Machida, Dr. Hideko Nomura, Dr. Nami Sakai, Dr. Masaki Sano, Dr. Yasushi Suto, Dr. Yoshimasa Watanabe, and all the members at the Yamamoto Group of Department of Physics, The University of Tokyo for their great support and discussion in various aspects of this research. Finally, the author thanks Prof. Takao Nakagawa, Prof. Aya Bamba, Prof. Hideko Nomura, Prof. Takashi Onaka, Prof. Masami Ouchi, and Prof. Jun'ichi Yokoyama for their valuable comments and suggestions as reviewers of this thesis. This thesis makes use of the following ALMA data sets: ADS/JAO.ALMA #2011.0.00604.S, ADS/JAO.ALMA #2011.0.00777.S, ADS/JAO.ALMA #2012.1.00712.S, ADS/JAO.ALMA #2013.1.01102.S, ADS/JAO. ALMA #2015.1.01060.S. ALMA is a partnership of the ESO (representing its member states), the NSF (USA), and NINS (Japan), together with the NRC (Canada) and the NSC and ASIAA (Taiwan), in cooperation with the Republic of Chile. The Joint ALMA Observatory is operated by the ESO, the AUI/NRAO, and the NAOJ. The author is grateful to the ALMA staff for their excellent support. The author also acknowledges the Advanced Leading Graduate Course for Photon Science (ALPS) and the JSPS fellowship for financial support. This study is supported by a Grant-in-Aid from the Ministry of Education, Culture, Sports, Science, and Technologies of Japan (15J01610).

Contents

Chapter 1
Introduction

1.1 Formation of Low-Mass (Solar-Type) Stars

Stars are born through gravitational collapse of dense parts of interstellar clouds, called molecular clouds. In the Milky Way galaxy, the total mass of molecular clouds is roughly 10^9 $M_\odot$, from which stars are continuously formed. The star-formation rate of molecular clouds in the solar neighborhood is 1.6 $M_\odot$ Myr^{-1} pc^{-2} [15], although it largely depends on molecular clouds.

In star formation studies, stars are roughly classified into two categories. Stars with a mass less than 8 $M_\odot$ are called as '*low-mass stars*' (or '*Solar-type stars*'), while ones with a higher mass are called as '*high-mass stars*'. This is because the time scale of star formation is much different between them. The timescale from the stellar birth to the main sequence is 10^{7-8} yr for low-mass stars, while it is as short as 10^5 yr for high-mass stars. High-mass stars often reach the main sequence before their parent cores are completely dissipated. Hence, the formation of high-mass stars heavily affects the environment of parent molecular clouds, and this fact results in a complexity of their formation process. The high-mass star formation study itself is an important and interesting subject. However, their observational studies often encounter various difficulties, because they are mostly distant from us. Although a rapid progress has been made by high angular-resolution observations for these years, the unified picture of high-mass star formation has not been established yet.

As for low-mass stars, especially isolated ones, their formation processes have extensively been studied both theoretically and observationally [e.g. 1, 5, 20, 21, 43, 60, 63]. Since planetary systems are potentially formed around low-mass stars, detailed understandings of their formation processes are essential to understanding the origin of the Solar System. So far, the following scenario is suggested as the formation of low-mass stars:

(1) Starless core
Before the birth of a protostar, parent molecular clouds are called as '*starless cores*'. While the gas collapses due to self-gravity, the temperature of the core is

Y. Oya, *A Few Tens au Scale Physical and Chemical Structures Around Young Low-Mass Protostars*, Springer Theses, https://doi.org/10.1007/978-981-19-1708-0_1

stably kept low (∼10 K) due to the radiative cooling (the isothermal collapse). Since the gas is usually sustained by the turbulence and the magnetic fields, the timescale of the starless-core stage is reported to be about $(0.5 \pm 0.3) \times 10^6$ yr, which is slightly longer than the free-fall time of the gas with the H_2 density of 10^4 cm^{-3} [15].

(2) Protostar

The central density of a starless core increases as the core collapses. When the central H_2 density exceeds 10^{11} cm^{-3}, the thermal radiation from dust grains is absorbed by the gas before it escapes from the core; namely, the core becomes '*optically thick*' for dust emission. This results in a low cooling efficiency, and thus the temperature in the central part of the core starts to rise. This optically thick core is called as '*first core*' [32]. Within 100 yr or so, the temperature of the core reaches 1500 K, and then the thermal dissociation of H_2 to 2H starts. Since the thermal dissociation of H_2 absorbs substantial energy, the gravitational collapse of the core is accelerated ('*second collapse*'). After this stage, the core is called as '*second core*', which is recognized as the birth of the protostar.
After the protostellar birth, the gas continues to accrete onto the protostar, and the protostar grows. The protostar shines by releasing the gravitational energy of the accreting gas ('*main accretion phase*'). Generally, the physical structure of the gas is no more spherically symmetric due to the angular momentum of the accreting gas, and the gas forms a flat envelope and a protostellar disk around the protostar. At this moment, a bipolar outflow perpendicular to the disk/envelope system also blows out from the protostar (e.g. [3]). It is generally thought that outflows are launched by the interaction between the gas motion and the magnetic field (e.g. [34]). They are thought to play an important role in extracting the angular momentum from the accreting gas. An outflow phenomenologically consists of an ionized jet and a molecular flow in its inner and outer parts, respectively. The velocity of an outflowing gas reaches as high as $(10 - 100)$ km s^{-1}. At this evolutionary stage, the gas simultaneously shows an infall motion in the envelope and an outflowing motion.

(3) T Tauri star

The envelope gas is dispersed by the outflow, and then the accretion of the gas onto the protostar gets settled ('*T Tauri star*'). At this evolutionary stage, the star shrinks quasi-statically so that the temperature in the central part rises. The heat generated inside the star is transported to its surface by convection ('*Hayashi phase*' [21]). When the temperature of the central part of the star becomes enough high, the heat is transported by radiation ('*Henyey track*' [22]). Then, T Tauri stars finally reach the main sequence in $(10^7 - 10^8)$ yr. Since the envelope gas is almost dissipated around a T Tauri star, the star is surrounded by only a protoplanetary disk, which is a parent of a planetary system. At the first phase of T Tauri stars ('*classical T Tauri stars*'), a small amount of the gas is still accreting onto the star through the protoplanetary disk, and it causes episodic phenomena, such as UV and X-ray flares (e.g. [17]). At the later phase of T Tauri stars, such phenomena due to the gas accretion scarcely occur, and the H_α line emission caused by the accretion is weakened ('*weak-line T Tauri stars*').

For stars before they reach the main sequence (‘*pre-main sequence stars*’), their radiation spectra in the infrared region (‘*spectral energy distribution*’; SED) are commonly used to classify their evolutionary stages phenomenologically [31]. Generally, the stellar evolutionary stages are classified into the stages called as ‘*Class I–III*’. [3] summarized the typical SEDs for these stages: Class I, II, and III corresponding to the evolutionary stages of the protostar, the classical T Tauri star, and the weak-line T Tauri star. Stars in earlier phases than Class I are called as Class 0 [2]. In these stages, the SED changes depending on which component in the source emits dominantly. In the Class 0 stage, the emitting source is a thick envelope gas at a low temperature which surrounds the protostar. It shows the black body radiation with its intensity peak in the far infrared region. Emission from the protostar is almost absorbed by the envelope, and its energy is reprocessed as far infrared emission. A Class I source is also surrounded by the envelope gas, and the radiation in the near infrared region is obscured and its peak is in the mid-infrared region. A Class II source shows the radiation from the stellar photosphere as well as that from the protoplanetary disk, and the radiation in the near infrared region is dominant. The SED of a Class III source is almost dominated by the black body radiation from the stellar photosphere in the near infrared region with a little excess in the infrared and UV regions. Since the SED is affected by various factors, such as the inclination angle of the protostellar/protoplanetary disk and the possible companion, this classification, especially that of Class 0 and I, has some uncertainty. Because of this reason, the bolometric temperature of sources is often employed for the classification; the bolometric temperature $T_{\rm bol}$ of an observed continuum spectrum is defined as the temperature of a black body having the same mean frequency as the observed spectrum [39]. Sources with the bolometric temperature less than 70 K are classified as Class 0, while ones with the temperature higher than 70 K are as Class I. In these classification of the evolutionary stages, the outflows blow during the stages from Class 0 to Class I. Protostellar disks are thought to be formed in the stages from Class 0 to Class I, while protoplanetary disks are in the stages from Class II to Class III. During the Class II and III stages, planetary systems are thought to be formed in protoplanetary disks. The life time scales for the starless-core, Class 0, I, II, and III stages are roughly 10^5, 10^4, 10^5, 10^6, and 10^7 yr, respectively [3].

1.2 Radio Astronomy

Radio wave is the electromagnetic wave, whose wavelength is longer than those of visible and infrared radiation. The boundary between the radio wave and the (far) infrared radiation is arbitary, which is often set at 100 μm (3 THz). The dawn of the radio astronomy was broken by Jansky in 1931, who detected a radio wave ($\lambda = 14.6$ m) from the Galactic Center [26]. This is emission of a huge plasma gas in interstellar medium. [46] observed the Milky Way with a hand-made parabola antenna, and mapped the distribution of the radio continuum emission ($\lambda = 1.85$ m). Then, [23] observed the Solar flare in radio wave, which is an explosive event on the Solar photosphere, producing a high temperature plasma gas and relativistic charged particles.

Radio observations have drastically been developed by making use of the development of the technologies of a microwave radar and communications during the World War II. [16, 37] discovered the 21 cm line of atomic hydrogen. In 1960s, Penzias and Wilson detected the cosmic microwave background emission, providing a definitive proof of the big-bang theory. New types of celestial objects, such as quasars and pulsers, were also found by radio observations. From the late 1960s to the early 1970s, spectral lines (mostly rotational spectral lines) of various molecular species, such as NH_3, H_2O, H_2CO, and CO, were found in the radio wave regions [12, 13, 62, 73], and moreover, '*interstellar molecular clouds*' were recognized as a dense and cold part of interstellar clouds, which harbor various molecular species. Since interstellar molecular clouds are the birthplaces of stars and planetary systems, they have extensively been studied as an important target for the radio astronomy. At the same time, detection of various interstellar molecules opened a new interdisciplinary field, '*astrochemistry*'.

Radio observations have several advantages in studies of star formation:

(1) Radio waves go through interstellar clouds with less attenuation due to the scattering and absorption by gas and dust than the electromagnetic wave at shorter wavelengths. Thus we can see through the central part of interstellar clouds, where protostars and protostellar disks are being formed. Moreover, the earth atmosphere is transparent in some frequency ranges of radio waves. These frequency ranges are called as '*radio window*' or '*atmospheric window*'. Hence, the radio waves in the windows can be observed by ground-based radio telescopes.
(2) Radio-wave photons have low energy, and can be emitted by cold matter. The black body radiation at a temperature from 10 to 50 K has its intensity peak in the millimeter/submillimeter region. The temperature of molecular clouds, which are the birthplaces of protostars, is typically as low as 10 K. Hence, their structures, including the protostellar envelope and disk system, can be traced by radio observations.
(3) The radio region contains various molecular lines. Rotational spectra of various molecular species, not only the basic species such as CO and CS but also more complex species such as $HCOOCH_3$ (methyl formate) and $(CH_3)_2O$ (dimethyl ether) fall in the radio-wave region, particularly in the millimeter and submillimeter wave regions. Thus, radio observations tell us what kinds of molecules are contained in the source and how much (Sect. 1.3).

On the other hand, radio astronomy has an apparent weakness: the angular resolution of radio telescopes is generally much poorer than those of visible and infrared telescopes with the same diameter (D). A diffraction-limited angular resolution of a telescope is proportional to $\frac{\lambda}{D}$, where λ is the wavelength. Thus, a radio telescope ($\lambda \sim 1\,\mathrm{mm}$) require a larger dish by 2000 times with respect to a visible telescope ($\lambda \sim 500\,\mathrm{nm}$) to achieve the same diffraction-limited angular resolution. However, the available size as a fully steerable dish is at most 100 m. Hence, radio astronomy overcomes this weakness by using the technique of the aperture synthesis, which is described in Chap. 2.

1.3 Astrochemistry in Star-Forming Region

Astrochemistry was initiated in the early 1970s thanks to the birth of millimeter-wave astronomy. Since then, breakthroughs in astrochemistry have always been triggered by new observational technologies. In this section, the progress of astrochemistry in star-forming regions is outlined in relation to technological developments.

1.3.1 1980s: Astrochemistry in High- and Low-Mass Star-Forming Region

Chemistry of star-forming regions was first studied for high-mass star-forming regions. This is simply because they have very bright emission of various molecular species. Therefore, their chemical compositions can readily be studied. Orion KL (Keinmann-Low object in Orion) is one of the best studied source in astrochemistry. It is known as the nearest high-mass star-forming region and the richest source in molecular emission (d = 437 pc [24]). In the early era of radio astronomy, the chemical composition of Orion KL was extensively studied by spectral line survey observations (e.g. [7, 27, 64, 71]) with the millimeter-wave telescopes (Onsala 20 m telescope, NRAO 12 m telescope, Owens Valley 10.4 m telescope). Later, the line survey of Orion KL has been expanded to higher frequencies [6, 56, 67]. The most characteristic chemical feature of Orion KL is the existence of various saturated complex organic molecules (COMs), such as $(CH_3)_2O$, $HCOOCH_3$, and C_2H_5CN. These species mostly resides in the hot and dense part near the protostar. Sulfur-containing molecules (SO and SO_2) and non-volatile molecules (SiO, SiS, and PN) are also found there. Chemical differentiation at a small spatial scale within Orion KL was inferred by the different spectral line shapes among molecules (e.g. [7]). This was indeed confirmed by interferometric observations (e.g. [35, 45, 74]). The N-bearing molecules are concentrated around '*hot core*' close to the high-mass protostar Source I, while the O-bearing molecules are distributed in a '*compact ridge*', which is a bit apart from the hot core. Observations toward high-mass protostellar sources have extensively been conducted to study its chemical composition itself and to search for new interstellar molecules.

In the 1980s, a number of low-mass protostars were identified in dense cores by the Infrared Astronomical Satellite (IRAS) (e.g. [4, 5]). However, chemical compositions of low-mass star-forming regions were relatively less studied in comparison with those of high-mass star-forming regions in those days. Since the molecular emission in low-mass star-forming regions is generally weaker than that in high-mass star-forming regions, less attention was paid for low-mass star-forming regions in astrochemistry. Some survey observations of only a few representative molecules, such as NH_3, CS, CCS, and HC_3N (e.g. [5, 65, 78]), were conducted toward low-mass protostellar sources. A systematic chemical difference was found between starless cores and star-forming cores [65]. However, this is a chemical evolutionary effect, and is not ascribed to the chemical composition in the vicinity of protostars (Fig. 1.1).

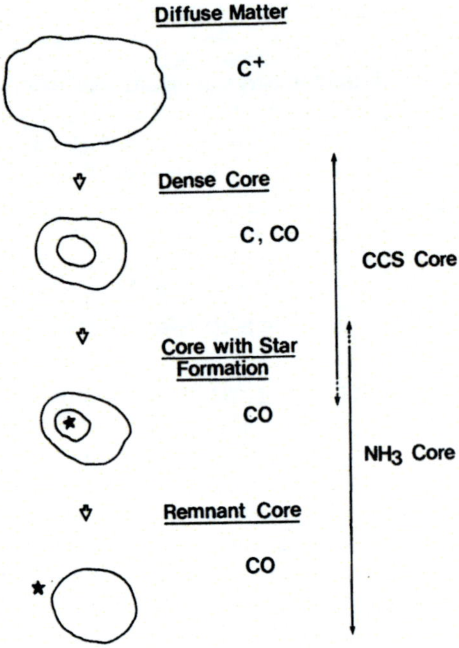

Fig. 1.1 Schematic illustration of the chemical evolution of a contracting cloud. Taken from [65]. © AAS. Reproduced with permission

1.3.2 1990s: Chemical Composition at a 10^3 au Scale

In the 1990s, the sensitivity of the radio telescope was much improved by using the superconducting heterodyne mixers, and the submillimeter-wave observations became possible. Thanks to these technological progresses, a pioneering study in astrochemistry of low-mass star-forming regions was conducted by [8, 72]. They carried out spectral line surveys in the millimeter/submillimeter-wave regions toward a low-mass protostellar source IRAS 16293−2422 in Ophiuchus (137 − 147 pc [44]) with spatial resolutions of $\sim 10^3$ au. They found that sulfur-containing molecules, including SO, SO_2, CS, and H_2CS, are abundant in this source. The emission of SO and SO_2 was thought to come from the hot and dense region near the protostar. Detection of high-excitation lines of CH_3OH clearly reveals the association of a 'hot core' like region in this source. However, the spectral lines of COMs, such as $HCOOCH_3$, $(CH_3)_2O$, and C_2H_5CN, were not detected in their observation, probably because of insufficient sensitivity. The existence of such a hot region was also inferred by the single-dish observations of H_2CO and CH_3OH toward this source and other low-mass protostellar sources (e.g. [11, 57]). With the aid of the simplified model considering the abundance jump in a hot region around the protostar, the size of the hot region was estimated to be as small as a few tens of au, although it was not able to be resolved by single-dish radio observations.

1.3.3 2000s: Chemical Diversity at a $10^2 - 10^3$ au Scale

The existence of the hot and dense region in IRAS 16293−2422 was confirmed in the 2000s. [10] detected various spectral lines of COMs, including $(CH_3)_2O$, $HCOOCH_3$, and C_2H_5CN, toward this source, and demonstrated the occurrence of a hot-core like chemistry even in a low-mass protostellar source. This discovery was made possible by development of sensitive receivers, whose noise temperature is close the quantum limit ($h\nu/k$). Such a hot and dense region around a low-mass protostar was called as '*hot corino*', named after its resemblance to the hot core in the high-mass star-forming region case with a much smaller scale. Indeed, interferometric studies of IRAS 16293−2422 show that those COMs exist in compact regions associated with the two components of the binary systems. It is thought that COMs are evaporated from grain mantle in the hot ($T \sim 100$ K) region around the protostar. Since then, some low-mass protostellar sources are also recognized to harbor a hot corino around their protostars (e.g. Serpens SMM1, SMM4, NGC1333 IRAS 2A, IRAS 4A, and IRAS 4B [9, 36, 40, 53]). The complex molecules detected toward hot corinos are thought to be related to the pre-solar organic materials found in meteorites, and thus, the discovery of hot corinos are of particular interest not only in astrochemistry but also in planetary science.

On the other hand, a low-mass protostellar source with a completely different chemical composition from IRAS 16293−2422 was found by [51]. They found that carbon-chain and related molecules, such as C_4H and c-C_3H_2, are abundant in a warm and dense region in the low-mass protostellar core L1527 in Taurus (137 pc [70]). Furthermore, they also detected some negatively charged carbon-chain molecules, such as C_6H^- and C_4H^- [52, 54]. In contrast to the IRAS 16293−2422 case, the spectral lines of $HCOOCH_3$ are not detected toward this source, and the CH_3OH lines are much weaker. The chemical characteristics found in L1527 is called as '*warm carbon-chain chemistry*' (WCCC [49, 55]). It is proposed that CH_4, which is an important constituent of grain mantle, is evaporated in a warm ($T \sim 30$ K) and dense region around the protostar, triggering efficient formation of various carbon-chain chemistry. [49] successively found another low-mass protostellar core having similar chemical characteristics to L1527, IRAS 15398−3359. The discovery of the WCCC sources established the chemical diversity in low-mass protostellar sources.

Since planetary systems are formed around low-mass stars, the chemical composition in low-mass star-forming regions, including its diversity, is essential to understanding the chemical evolution from molecular clouds to planetary systems.

1.3.4 2010s: ALMA Era at $10 - 10^2$ au Scale

With the advent of Atacama Large Millimeter/submillimeter Array (ALMA), astrochemistry in low-mass star-forming regions have extensively been advanced. ALMA has allowed us to study thermal emission of less abundant molecules at a subarcsecond resolution. One of the most notable studies conducted with ALMA is the

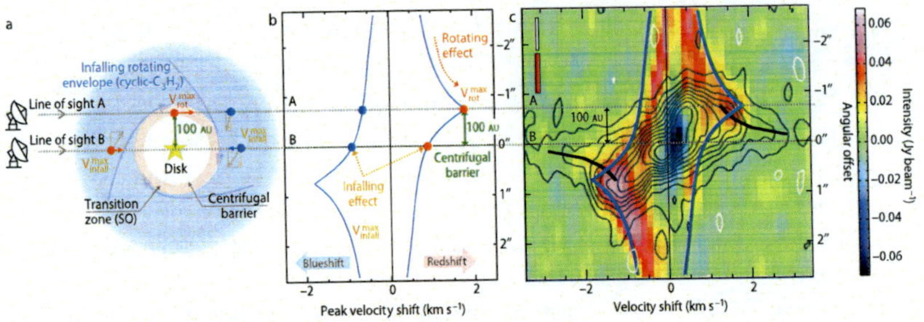

Fig. 1.2 **a** Schematic illustration of the infalling-rotating envelope. **b** Maximum velocity-shifts calculated with the model for the edge-on configuration. **c** Position-velocity diagrams of c-C_3H_2 (color) and SO (contours) along the mid-plane of the disk/envelope system in L1527. Blue and black lines represent the maximum velocity-shifts of the infalling-rotating and Keplerian motion, respectively, calculated by using the models explained in Chap. 3. Taken from [50]. Reprinted by permission from *Nature* 507(7490):78–80 (2014)

discovery of the centrifugal barrier of an infalling-rotating envelope around the protostar in the WCCC source L1527 (Fig. 1.2 [50]). They found that the CCH emission in the envelope gas abruptly disappears at the distance of 100 au from the protostar. Before the ALMA era, kinematic structures of the gas around protostars have often been interpreted as the Keplerian motion, where the gas is rotationally supported, without careful considerations. Although the infall motion in L1527 was inferred in the analysis of the CO isotopologue lines at a resolution of ~1000 au (e.g. [41]), its detailed characterization was difficult because of the limited resolution and sensitivity. With ALMA, the kinematic structure traced by CCH in L1527 was confirmed to have infall motion, and it was simply explained by the ballistic motion. In the infalling-rotating envelope, the gas cannot fall inward of a certain radius, because of the energy and angular momentum conservation. This position corresponds to the 'perihelion', and is called as '*centrifugal barrier*'. [50] characterized the physical parameters of its infalling-rotating envelope at a 100 au scale, and identified its centrifugal barrier. They also reported a drastic chemical change across the centrifugal barrier in L1527 [48, 50]. The CCH and c-C_3H_2 lines trace the infalling-rotating envelope, while the SO line selectively highlights the centrifugal barrier. Since the radius of the centrifugal barrier (100 au) corresponds to the typical size of a protoplanetary disk system, such a chemical differentiation around the centrifugal barrier is an important clue to understand what kinds of molecules are delivered into the disk component. We are now going into the era of observations at the disk-forming region scale with ALMA.

1.4 Motivation for This Research

The formation process of low-mass (Solar-type) stars and their planetary systems is classified into the following three stages: (1) the contraction of the interstellar molecular cloud and the birth of the protostar inside it; (2) the formation of the protostellar/protoplanetary disk around the protostar; (3) the formation of the planetary system in the disk. As described above, radio and infrared observations have greatly enhanced our understandings of the first and the last stages of the formation process recently. However, the formation processes of the protostellar/protoplanetary disk and the associated material evolution are still unknown. Therefore, understanding the formation processes of rotationally-supported disks around young low-mass protostars is an important target for star formation studies. Rotationally-supported disks are usually found around low-mass Class I protostars (e.g. [25, 66, 76]), and a few observational evidences of disks associated with the Class 0 stage have also been reported (e.g. [38, 42, 68, 75, 77]). Hence, it seems likely that the disk structure is formed in an early stage of protostellar evolution.

The rotation velocity of the Keplerian motion is proportional to $r^{-0.5}$, where r denotes the radius from the protostar. On the other hand, the rotation velocity of the infalling-rotating motion conserving the angular momentum shows r^{-1} dependence. Based on these relations, the disk radius is evaluated as the knee point of the radial dependence of the velocity centroid (e.g., Fig. 1.3 [42]). However, it should be noted that the evaluation of the disk radius may suffer from the complex structure of the disk/envelope system, if it is not spatially resolved.

Disk formation processes are deeply related to angular momentum of the infalling gas of the envelope. In these early phases of protostellar evolution, energetic outflows blow from protostars, which means that accretion and ejection of mass are occurring at the same time (e.g. [3]). Outflows are thought to play an important role in extracting angular momentum of the infalling gas (e.g. [19, 34, 59, 61, 69]). Hence,

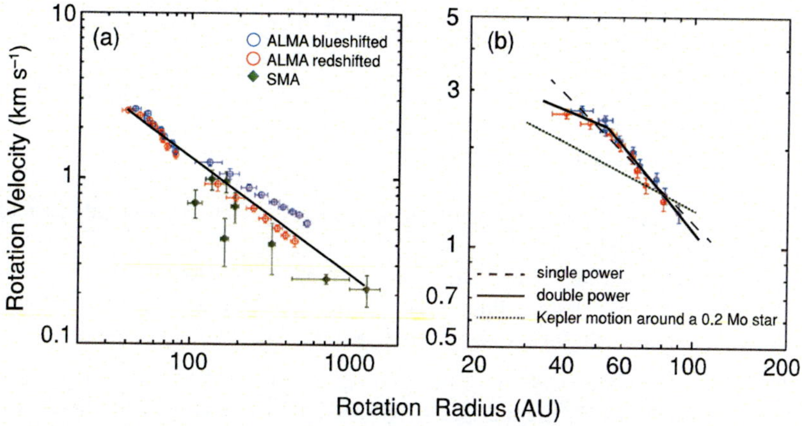

Fig. 1.3 Radial dependence of the rotation velocity traced by the $C^{18}O$ line. **a** Blue and red marks represent the data points at the blue- and red-shifted velocities, respectively. Green marks represent the data points observed with SMA [77] for comparison. **b** Blow-up of panel (**a**), where only the data points within the radius of 100 au are shown. The dashed and solid lines represent the least-square fitting with single and two power laws, respectively. The dotted line represents the radial dependence of the Keplerian velocity with the protostellar mass of 0.2 $M_{\odot}$. Taken from [42].

both disk formation and outflow launching are related to angular momentum, and understanding one of the two would help us to understand the other.

The existence of the centrifugal barrier of the infalling-rotating envelope and the associated drastic chemical change mentioned above were totally unexpected in star-formation studies and astrochemical studies, and their discovery opened a new door toward understanding physical and chemical processes in disk formation and outflow-launching. I started my PhD work just after the discovery of the centrifugal barrier in L1527.

For better understandings of the observed kinematic structure of the infalling-rotating envelope in L1527, I constructed a simplified three dimensional model assuming the ballistic motion of the gas, and simulated the position-velocity (PV) diagram shown in Fig. 1.2. As demonstrated later in Chap. 3, the model successfully reproduces the kinematic structure of the infalling-rotating envelope. From this result, it is confirmed that CCH and c-C_3H_2 preferentially exist in the infalling-rotating envelope. With the aid of this model, the protostellar mass and the radius of the centrifugal barrier of L1527 are evaluated to be 0.18 $M_\odot$ and 100 au, respectively. This exercise gave me a strong impression that the model, despite simplified one, can be a powerful tool to extract the physical parameters of the infalling-rotating envelope.

Although the basic concept of the centrifugal barrier had been reported for L1527 by [50], important new questions related to the centrifugal barrier were raised to be addressed. Several of them are:

(1) Is the centrifugal barrier of the infalling-rotating envelope a common occurrence in low-mass protostellar sources?
(2) What is a role of the centrifugal barrier in star formation? How is the disk structure formed inside the centrifugal barrier? Is there any relation between the centrifugal barrier and the outflow launching?
(3) How does the magnetic field affect the centrifugal barrier and the disk formation processes?
(4) What a chemical change occurs at the centrifugal barrier? Is the chemical change different between the WCCC sources and the hot corino sources?
(5) What is the chemical heritage passed from the infalling-rotating envelope to the disk component?
(6) Are there molecular lines that can selectively trace specific physical components (i.e., the infalling-rotating envelope, the centrifugal barrier, and the disk component) for the chemical diagnostics?

To answer these questions, observations of various low-mass protostellar sources other than L1527 are indispensable. In astrophysical and astrochemical studies, we cannot artificially control the environments of the target source unlike the laboratory experiments. It is also impossible to trace the evolution of the target source, because the time scale of the astrophysical and astrochemical phenomena is always quite longer than the lifetime of human beings. Instead, we can investigate the environmental effect and the evolutionary effect by observing a number of sources. Such

Table 1.1 Target sources studied in this thesis

Source Name	Distance (pc)	Luminosity ($L_\odot$)	Chemistry
L1527	137[a]	1.7[b]	WCCC[c]
IRAS 15398−3359	155[d]	1.8[e]	WCCC[f]
IRAS 16293−2422 Source A	137 − 147[g]	22[h]	Hot Corino[i]
IRAS 16293−2422 Source B	137 − 147[g]	22[h]	Hot Corino[i]
L483	200[j]	13[k]	Hybrid[l]

[a]Taken from [70]
[b]Taken from [18]
[c]e.g. [51]. WCCC stands for the warm carbon-chain chemistry
[d]Taken from [33]
[e]Taken from [29]
[f]Taken from [49]
[g]Taken from [44]
[h]The value for the binary system IRAS 16293−2422 taken from [14], assuming *d* of 120 pc [30]
[i]e.g. [10, 57]
[j]Taken from [28, 47]
[k]Taken from [58]
[l]See Chap. 8

an approach is standard in astrophysics and astrochemistry. With this in mind, I observationally investigated the physical and chemical structures of five low-mass protostellar sources at a few 10 au scale by using ALMA, as the initial step toward statistical studies. The sources studied in this thesis are listed in Table 1.1. The list involves two WCCC sources (L1527 and IRAS 15398−3359), two hot corino sources (IRAS 16293−2422 Source A and Source B), and one hybrid (intermediate) character source (L483). They are all representative low-mass protostellar sources in the Solar neighborhood. By analyzing the observational results with the aid of the kinematic model, the physical and chemical processes occurring in disk formation will be investigated in various protostellar sources. This is the principal aim of this thesis. It will contribute to bridging the missing link in the low-mass star formation process mentioned above.

1.5 Outline of This Thesis

This thesis consists of 11 chapters. After this introductory chapter, the mechanisms of interferometers and the principles of aperture synthesis are described (Chap. 2). Chapter 3 deals with the kinematic models of the infalling-rotating envelope, the Keplerian disk, and the outflow. In addition to the details of the models, some examples of the model simulations are presented. Chapters 4−8 present the observational results Chaps. 4−8 and analyses for the individual protostellar sources listed in Table 1.1. The overall discussions based on the observational results are given in Chaps. 9 and 10 for chemical and physical implications, respectively. Finally, Chap. 11 concludes the thesis with a few future prospects.

References

1. André Ph, Men'shchikov A, Bontemps S, Könyves V, Motte F, Schneider N, Didelon P, Minier V, Saraceno P, Ward-Thompson D, di Francesco J, White G, Molinari S, Testi L, Abergel A, Griffin M, Henning Th, Royer P, Merín B, Vavrek R, Attard M, Arzoumanian D, Wilson CD, Ade P, Aussel H, Baluteau JP, Benedettini M, Bernard JPh, Blommaert JADL, Cambrésy L, Cox P, di Giorgio A, Hargrave P, Hennemann M, Huang M, Kirk J, Krause O, Launhardt R, Leeks S, Le Pennec J, Li JZ, Martin PG, Maury A, Olofsson G, Omont A, Peretto N, Pezzuto S, Prusti T, Roussel H, Russeil D, Sauvage M, Sibthorpe B, Sicilia-Aguilar A, Spinoglio L, Waelkens C, Woodcraft A, Zavagno A (2010) From filamentary clouds to prestellar cores to the stellar IMF: Initial highlights from the Herschel Gould Belt Survey. A&A 518:L102
2. Andre P, Ward-Thompson D, Barsony M (1993) Submillimeter Continuum Observations of rho Ophiuchi A: the Candidate Protostar VLA 1623 and Prestellar Clumps. ApJ 406:122
3. Bachiller R (1996) Bipolar molecular outflows from young stars and protostars. ARA&A 34:111–154
4. Beichman CA, Myers PC, Emerson JP, Harris S, Mathieu R, Benson PJ, Jennings RE (1986) Candidate solar-type protostars in nearby molecular cloud cores. ApJ 307:337
5. Benson PJ, Myers PC (1989) A survey for dense cores in dark clouds. ApJS 71:89
6. Beuther H, Zhang Q, Greenhill LJ, Reid MJ, Wilner D, Keto E, Shinnaga H, Ho PTP, Moran JM, Liu SY, Chang CM (2005) Line imaging of orion KL at 865 μm with the submillimeter array. ApJ 632(1):355–370
7. Blake GA, Sutton EC, Masson CR, Phillips TG (1987) Molecular abundances in OMC-1: the chemical composition of interstellar molecular clouds and the influence of massive star formation. ApJ 315:621
8. Blake GA, van Dishoeck EF, Jansen DJ, Groesbeck TD, Mundy LG (1994) Molecular abundances and low-mass star formation. I. Si- and S-bearing species toward IRAS 16293–2422. ApJ 428:680
9. Bottinelli S, Ceccarelli C, Lefloch B, Williams JP, Castets A, Caux E, Cazaux S, Maret S, Parise B, Tielens AGGM (2004) Complex molecules in the hot core of the low-mass protostar NGC 1333 IRAS 4A. ApJ 615(1):354–358
10. Cazaux S, Tielens AGGM, Ceccarelli C, Castets A, Wakelam V, Caux E, Parise B, Teyssier D (2003) The hot core around the low-mass protostar IRAS 16293–2422: Scoundrels Rule! ApJ 593(1):L51–L55
11. Ceccarelli C, Castets A, Loinard L, Caux E, Tielens AGGM (1998) Detection of doubly deuterated formaldehyde towards the low-luminosity protostar IRAS 16293–2422. A&A 338:L43–L46
12. Cheung AC, Rank DM, Townes CH, Thornton DD, Welch WJ (1968) Detection of NH_3 molecules in the interstellar medium by their microwave emission. Phys Rev Lett 21(25):1701–1705
13. Cheung AC, Rank DM, Townes CH, Thornton DD, Welch WJ (1969) Detection of water in interstellar regions by its microwave radiation. Nature 221(5181):626–628
14. Crimier N, Ceccarelli C, Maret S, Bottinelli S, Caux E, Kahane C, Lis DC, Olofsson J (2010) The solar type protostar IRAS16293-2422: new constraints on the physical structure. A&A 519:A65
15. II Evans NJ, Dunham MM, Jørgensen JK, Enoch ML, Merín B, Van Dishoeck EF, Alcalá JM, Myers PC, Stapelfeldt KR, Huard TL, Allen LE, Harvey PM, van Kempen T, Blake GA, Koerner DW, Mundy LG, Padgett DL, Sargent AI (2009) The spitzer c2d legacy results: star-formation rates and efficiencies; evolution and lifetimes. ApJS, 181(2):321–350
16. Ewen HI, Purcell EM (1951) Observation of a line in the galactic radio spectrum: radiation from galactic hydrogen at 1,420 Mc./s. Nature 168(4270):356
17. Feigelson Eric D, Montmerle Thierry (1999) High-energy processes in young stellar objects. ARA&A 37:363–408

18. Green JD, II Evans NJ, Jørgensen JK, Herczeg GJ, Kristensen LE, Lee JE, Dionatos O, Yildiz UA, Salyk C, Meeus G, Bouwman J, Visser R, Bergin EA, van Dishoeck EF, Rascati MR, Karska A, van Kempen TA, Dunham MM, Lindberg JE, Fedele D, DIGIT Team (2013) Embedded protostars in the dust, ice, and gas in time (DIGIT) herschel key program: continuum SEDs, and an inventory of characteristic far-infrared lines from PACS spectroscopy. ApJ 770(2):123
19. Hartmann L (2009) The star-jet-disk system and angular momentum transfer. Astrophys Space Sci Proc 13:23–32
20. Hartmann L, Herczeg G, Calvet N (2016) Accretion onto pre-main-sequence stars. ARA&A 54:135–180
21. Hayashi C (1961) Stellar evolution in early phases of gravitational contraction. PASJ 13:450–452
22. Henyey LG, Lelevier R, Levée RD (1955) The early phases of stellar evolution. PASP 67(396):154
23. Hey JS (1946) Solar radiations in the 4–6 metre radio wave-length band. Nature 157(3976):47–48
24. Hirota T, Bushimata T, Choi YK, Honma M, Imai H, Iwadate K, Jike T, Kameno S, Kameya O, Kamohara R, Kan-Ya Y, Kawaguchi N, Kijima M, Kim MK, Kobayashi H, Kuji S, Kurayama T, Manabe S, Maruyama K, Matsui M, Matsumoto N, Miyaji T, Nagayama T, Nakagawa A, Nakamura K, Sik Oh C, Omodaka T, Oyama T, Sakai S, Sasao T, Sato K, Sato M, Shibata KM, Shintani M, Tamura Y, Tsushima M, Yamashita K (2007) Distance to Orion KL measured with VERA. PASJ 59:897
25. Hogerheijde Michiel R (2001) From Infall to Rotation around young stellar objects: a transitional phase with a 2000 AU radius contracting disk? ApJ 553(2):618–632
26. Jansky Karl G (1933) Radio waves from outside the solar system. Nature 132(3323):66
27. Johansson LEB, Andersson C, Ellder J, Friberg P, Hjalmarson A, Hoglund B, Irvine WM, Olofsson H, Rydbeck G (1984) Spectral scan of Orion A and IRC +10216 from 72 to 91 GHz. A&A 130:227–256
28. Jørgensen JK, Schöier FL, van Dishoeck EF (2002) Physical structure and CO abundance of low-mass protostellar envelopes. A&A 389:908–930
29. Jørgensen Jes K, Visser Ruud, Sakai Nami, Bergin Edwin A, Brinch Christian, Harsono Daniel, Lindberg Johan E, van Dishoeck Ewine F, Yamamoto Satoshi, Bisschop Suzanne E, Persson Magnus V (2013) A recent accretion burst in the low-mass protostar IRAS 15398–3359: ALMA imaging of its related chemistry. ApJ 779(2):L22
30. Knude J, Hog E (1998) Interstellar reddening from the HIPPARCOS and TYCHO catalogues. I. Distances to nearby molecular clouds and star forming regions. A&A 338:897–904
31. Lada CJ, Wilking BA (1984) The nature of the embedded population in the rho Ophiuchi dark cloud?: mid-infrared observations. ApJ 287:610–621
32. Larson RB (1969) Numerical calculations of the dynamics of collapsing proto-star. MNRAS 145:271
33. Lombardi M, Lada CJ, Alves J (2008) Hipparcos distance estimates of the Ophiuchus and the Lupus cloud complexes. A&A 480(3):785–792
34. Machida MN, Hosokawa T (2013) Evolution of protostellar outflow around low-mass protostar. MNRAS 431(2):1719–1744
35. Masson Colin R, Mundy Lee G (1988) The hot core of Orion: 4 Maps of HC 3N emission. ApJ 324:538
36. Maury AJ, Belloche A, André P, Maret S, Gueth F, Codella C, Cabrit S, Testi L, Bontemps S (2014) First results from the CALYPSO IRAM-PdBI survey. II. Resolving the hot corino in the Class 0 protostar NGC 1333-IRAS2A. A&A, 563:L2
37. Muller CA, Oort JH (1951) Observation of a line in the galactic radio spectrum: the interstellar hydrogen line at 1,420 Mc./s, and an estimate of galactic rotation. Nature, 168(4270):357–358
38. Murillo NM, Lai S-P, Bruderer S, Harsono D, van Dishoeck EF (2013) A Keplerian disk around a Class 0 source: ALMA observations of VLA1623A. A&A 560:A103
39. Myers PC, Ladd EF (1993) Bolometric temperatures of young stellar objects. ApJ 413:L47

40. Öberg Karin I, van der Marel Nienke, Kristensen Lars E, van Dishoeck Ewine F (2011) Complex molecules toward low-mass protostars: the serpens core. ApJ 740(1):14
41. Ohashi N, Hayashi M, Ho PTP, Momose M (1997) Interferometric imaging of IRAS 04368+2557 in the L1527 molecular cloud core: a dynamically infalling envelope with rotation. ApJ, 475(1):211–223
42. Ohashi N, Saigo K, Aso Y, Aikawa Y, Koyamatsu S, Machida MN, Saito M, Takahashi SZ, Takakuwa S, Tomida K, Tomisaka K, Yen H-W (2014) Formation of a Keplerian disk in the infalling envelope around L1527 IRS: transformation from infalling motions to Kepler motions. ApJ 796(2):131
43. Onishi T, Mizuno A, Kawamura A, Ogawa H, Fukui Y (1996) A C 18O Survey of dense cloud cores in taurus: core properties. ApJ 465:815
44. Ortiz- León GN, Loinard L, Kounkel MA, Dzib SA, Mioduszewski AJ, Rodríguez LF, Torres RM, González-Lópezlira RA, Pech G, Rivera JL, Hartmann L, Boden AF, II Evans NJ, Briceño C, Tobin JJ, Galli PAB, Gudehus D (2017) The Gould's Belt Distances Survey (GOBELINS). I. Trigonometric parallax distances and depth of the ophiuchus complex. ApJ, 834(2):141
45. Plambeck RL, Wright MCH (1987) Aperture synthesis maps of HDO emission in Orion-KL. ApJ 317:L101
46. Reber G (1940) Notes: cosmic static. ApJ 91:621–624
47. Rice EL, Prato L, McLean IS (2006) An association in the aquila star-forming region: high-resolution infrared spectroscopy of T Tauri Stars. ApJ 647(1):432–443
48. Sakai N, Oya Y, Sakai T, Watanabe Y, Hirota T, Ceccarelli C, Kahane C, Lopez-Sepulcre A, Lefloch B, Vastel C, Bottinelli S, Caux E, Coutens A, Aikawa Y, Takakuwa S, Ohashi N, Yen H-W, Yamamoto S (2014) A chemical view of protostellar-disk formation in L1527. ApJ 791(2):L38
49. Sakai N, Sakai T, Hirota T, Burton M, Yamamoto S (2009) Discovery of the second warm carbon-chain-chemistry source, IRAS15398–3359 in Lupus. ApJ 697(1):769–786
50. Sakai N, Sakai T, Hirota T, Watanabe Y, Ceccarelli C, Kahane C, Bottinelli S, Caux E, Demyk K, Vastel C, Coutens A, Taquet V, Ohashi N, Takakuwa S, Yen H-W, Aikawa Y, Yamamoto S (2014) Change in the chemical composition of infalling gas forming a disk around a protostar. Nature 507(7490):78–80
51. Sakai N, Sakai T, Hirota T, Yamamoto S (2008) Abundant carbon-chain molecules toward the low-mass protostar IRAS 04368+2557 in L1527. ApJ 672(1):371–381
52. Sakai N, Sakai T, Osamura Y, Yamamoto S (2007) Detection of C_6H^- toward the low-mass protostar IRAS 04368+2557 in L1527. ApJ 667(1):L65–L68
53. Sakai N, Sakai T, Yamamoto S (2006) Detection of $HCOOCH_3$ toward a low-mass protostar, NGC 1333 IRAS 4B. PASJ 58:L15–L18
54. Sakai N, Sakai T, Yamamoto S (2008) Complex organic molecules in an early stage of protostellar evolution. Ap&SS 313(1–3):153–157
55. Sakai N, Yamamoto S (2013) Warm carbon-chain chemistry. Chem Rev 113(12):8981–9015
56. Schilke P, Groesbeck TD, Blake GA, Phillips TG (1997) A line survey of Orion KL from 325 to 360 GHz. ApJS 108(1):301–337
57. Schöier FL, Jørgensen JK, van Dishoeck EF, Blake GA (2002) Does IRAS 16293–2422 have a hot core? Chemical inventory and abundance changes in its protostellar environment. A&A 390:1001–1021
58. Shirley YL, II Evans NJ, Rawlings JMC, Gregersen EM (2000) Tracing the mass during low-mass star formation. I. Submillimeter continuum observations. ApJS 131(1):249–271
59. Shu F, Najita J, Ostriker E, Wilkin F, Ruden S, Lizano S (1994) Magnetocentrifugally driven flows from young stars and disks. I. A generalized model. ApJ 429:781
60. Shu FH, Adams FC, Lizano S (1987) Star formation in molecular clouds: observation and theory. ARA&A 25:23–81
61. Shu FH, Najita J, Ruden SP, Lizano S (1994) Magnetocentrifugally driven flows from young stars and disks. II. Formulation of the dynamical problem. ApJ, 429:797
62. Snyder LE, Buhl D, Zuckerman B, Palmer P (1969) Microwave detection of interstellar formaldehyde. Phys Rev Lett 22(13):679–681

63. Stahler SW, Shu FH, Taam RE (1980) The evolution of protostars. I—Global formulation and results. ApJ 241:637–654
64. Sutton EC, Blake GA, Masson CR, Phillips TG (1985) Molecular line survey of Orion A from 215 to 247 GHz. ApJS 58:341–378
65. Suzuki H, Yamamoto S, Ohishi M, Kaifu N, Ishikawa S-I, Hirahara Y, Takano S (1992) A survey of CCS, HC 3N, HC 5N, and NH 3 toward dark cloud cores and their production chemistry. ApJ 392:551
66. Takakuwa S, Saito M, Lim J, Saigo K, Sridharan TK, Patel NAl (2012) A Keplerian circumbinary disk around the protostellar system L1551 NE. ApJ 754(1):52
67. Tercero B, Cernicharo J, Pardo JR, Goicoechea JR (2010) A line confusion limited millimeter survey of Orion KL. I. Sulfur carbon chains. A&A 517:A96
68. Tobin JJ, Hartmann L, Chiang H-F, Wilner DJ, Looney LW, Loinard L, Calvet N, D'Alessio P (2012) A $\sum$0.2-solar-mass protostar with a Keplerian disk in the very young L1527 IRS system. Nature 492(7427):83–85
69. Tomisaka K (2002) Collapse of rotating magnetized molecular cloud cores and mass outflows. ApJ 575(1):306–326
70. Torres RM, Loinard L, Mioduszewski AJ, Rodríguez LF (2007) VLBA Determination of the distance to nearby star-forming regions. II. Hubble 4 and HDE 283572 in Taurus. ApJ 671(2):1813–1819
71. Turner BE (1989) A molecular line survey of sagittarius B2 and Orion-KL from 70 to 115 GHz. I. The observational data. ApJS 70:539
72. van Dishoeck EF, Blake GA, Jansen DJ, Groesbeck TD (1995) Molecular abundances and low-mass star formation. II. Organic and deuterated species toward IRAS 16293-2422. ApJ 447:760
73. Wilson RW, Jefferts KB, Penzias AA (1970) Carbon monoxide in the Orion Nebula. ApJ 161:L43
74. Wilson TL, Johnston KJ, Henkel C, Menten KM (1989) The distribution of hot thermal methanol in Orion-KL. A&A 214:321–326
75. Yen H-W, Koch PM, Takakuwa S, Krasnopolsky R, Ohashi N, Aso Y (2017) Signs of early-stage disk growth revealed with ALMA. ApJ 834(2):178
76. Yen H-W, Takakuwa S, Ohashi N, Aikawa Y, Aso Y, Koyamatsu S, Machida MN, Saigo K, Saito M, Tomida K, Tomisaka K (2014) ALMA observations of infalling flows toward the Keplerian disk around the Class I Protostar L1489 IRS. ApJ 793(1):1
77. Yen H-W, Takakuwa S, Ohashi N, Ho PTP (2013) Unveiling the evolutionary sequence from infalling envelopes to keplerian disks around low-mass protostars. ApJ 772(1):22
78. Zhou S, Wu Y, II Evans NJ, Fuller GA, Myers PC (1989) A CS survey of low-mass cores and comparison with NH 3 observations. ApJ 346:168

Chapter 2
ALMA Observation

2.1 Principles of Interferometers

ALMA employs a technique of the aperture synthesis, mixing signals from a cluster of antennas, to obtain a radiation source image at a high angular resolution. This technique derives the image of the source from the difference of the phase between the signals detected by different antennas. Ryle established the technique, and received a Noble Prize for this technique conjointly with Hewish for the other contribution.

2.1.1 *Coordinate System*

Here, **s** is defined as a unit vector with the direction from the earth to the radiation source. The origin is taken at the center of the radiation source on the plane of the sky. l and m axes are defined along the east-west and south-north direction, respectively, through the origin on the plane of the sky (Fig. 2.1). Hereafter, the radiation source is assumed to be distributed within a compact region where $|l|$ and $|m|$ can be assumed to be infinitesimal, and thus the curvature of the plane of the sky is ignored. The vector from one antenna to the other is called as a *'baseline vector'*. Thus a cluster of n antennas has $n\ (n-1)$ baseline vectors. For two antennas with a baseline vector **D**, the radiation from the source is detected by them with a mutual time delay (τ_g) of $\frac{1}{c}\ (\mathbf{D} \cdot \mathbf{s})$, where c denotes the speed of light. Here, τ_g is called as a *'geometrical delay'* (Fig. 2.2).

2.1.2 *Correlation Function and Cross Power Spectrum*

If the radiation from the source is monochromatic, which is represented as $E\ (t) = E_0 \exp\ (2\pi i \nu_0 t)$, the signals detected by two antennas are represented as:

Y. Oya, *A Few Tens au Scale Physical and Chemical Structures Around Young Low-Mass Protostars*, Springer Theses, https://doi.org/10.1007/978-981-19-1708-0_2

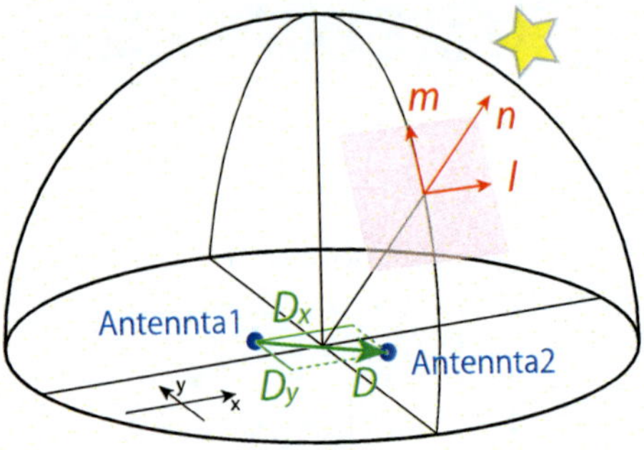

Fig. 2.1 Coordinate system of the plane of the sky and a baseline vector (**D**)

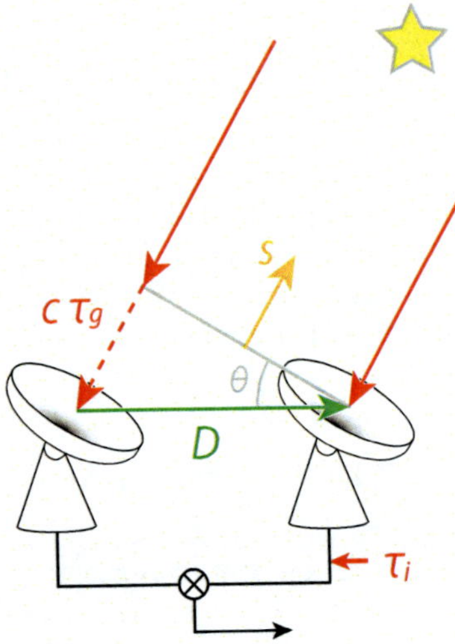

Fig. 2.2 Geometrical delay (τ_g) for two antennas

$$\tilde{E}_1(t) = a_1 E\,(t)\,, \quad \tilde{E}_2(t) = a_2 E\left(t - \tau_g\right), \tag{2.1}$$

where a_i is the complex gain of the antenna i. When a correction of the delay (τ_i) is applied to the antenna 1, the correlation function of the two signals as a function of a delay time τ is represented as:

$$\begin{aligned} C(\tau) &= \lim_{T\to\infty} \frac{1}{T} \int_{-T/2}^{T/2} \tilde{E}_1(t-\tau_i)\tilde{E}_2^*(t-\tau)dt \\ &= \lim_{T\to\infty} \frac{1}{T} \int_{-T/2}^{T/2} a_1 a_2^* |E_0|^2 \exp(2\pi i \nu_0((t-\tau_i)-(t-\tau_g-\tau)))dt \\ &= a_1 a_2^* |E_0|^2 \exp(2\pi i \nu_0(\tau+\tau_g-\tau_i)). \end{aligned} \tag{2.2}$$

$C(\tau)$ is the output data from the interferometer, and the cross power spectrum ($\hat{C}(\nu)$) is obtained by its Fourier transformation as:

$$\begin{aligned}\hat{C}(\nu) &= \int_{-\infty}^{\infty} C(\tau)\exp(-2\pi i\nu\tau)d\tau \\ &= a_1 a_2^* |E_0|^2 \delta(\nu_0 - \nu)\exp(2\pi i\nu_0(\tau_g - \tau_i)).\end{aligned} \tag{2.3}$$

When τ_i is correctly taken to be τ_g, the phase of $\hat{C}(\nu)$ $\left(2\pi i\nu_0\left(\tau_g - \tau_i\right)\right)$ equals to 0. Hence, the cross power spectrum for a monochromatic radiation at a frequency of ν_0 has a non-zero value only for ν of ν_0.

The radiation from the source is not always monochromatic. Even in this case, it can be approximated to the quasi-monochromatic radiation represented as:

$$\begin{aligned}E(t) &= \int_{-\infty}^{\infty} \hat{E}(\nu)\exp\left(2\pi i\nu t\right)d\nu \\ &= a_m(t)\exp\left(2\pi i\nu_0 t\right),\end{aligned} \tag{2.4}$$

where $a_m(t)$ denotes the time dependent amplitude of the radiation. Here, the correlation function (Eq. 2.2) is rewritten as:

$$\begin{aligned}C(\tau) &= \lim_{T\to\infty}\frac{1}{T}\int_{-T/2}^{T/2} \tilde{E}_1(t-\tau_i)\tilde{E}_2^*(t-\tau)dt \\ &= \lim_{T\to\infty}\frac{1}{T}\int_{-T/2}^{T/2} a_1 a_2^* a_m(t-\tau_i)a_m^*(t-\tau_g-\tau)\exp(2\pi i\nu_0(\tau+\tau_g-\tau_i))dt \\ &= a_1 a_2^* < a_m(t)a_m^*(t) >_T \exp(2\pi i\nu_0(\tau+\tau_g-\tau_i)).\end{aligned} \tag{2.5}$$

Here, the temporal variation of $a_m(t)$ is assumed to be sufficiently longer than $(\tau + \tau_g - \tau_i)$, so that $\lim_{T\to\infty}\frac{1}{T}\int_{-T/2}^{T/2} a_m(t-\tau_i)a_m^*(t-\tau_g-\tau)dt$ can be approximated to be the time-averaged power spectrum of the radiation $< a_m(t)a_m^*(t) >_T (= |E_0|^2)$. Then, the cross power spectrum is obtained as:

$$\begin{aligned}\hat{C}(\nu) &= \int_{-\infty}^{\infty} C(\tau)\exp(-2\pi i\nu\tau)d\tau \\ &= a_1 a_2^* < a_m(t)a_m^*(t) >_T \delta(\nu_0 - \nu)\exp(2\pi i\nu_0(\tau_g - \tau_i)).\end{aligned} \tag{2.6}$$

Again, the phase of $\hat{C}(\nu)$ equals to 0 when τ_i is correctly taken to be τ_g. Thus, the radiation can be treated as if it were monochromatic.

2.1.3 Visibility

For simplicity, one-dimensional distribution of the source on the l axis is first considered, and then the result is expanded to the two-dimensional distribution later. With an angular offset of Δl for the position of the radiation source, the phase of $\hat{C}(\nu)$ $\left(\phi = 2\pi\nu(\tau_g - \tau_i)\right)$ changes by $\Delta\phi$. $\Delta\phi$ is represented as:

$$\begin{aligned} \Delta\phi &= \frac{\partial\phi}{\partial l}\Delta l \\ &= 2\pi\nu\frac{\partial\tau_g}{\partial l}\Delta l \\ &= \frac{2\pi\nu}{c}\frac{\partial}{\partial l}(\mathbf{D}\cdot\mathbf{s})\Delta l \\ &= \frac{2\pi}{\lambda}D\cos\theta\Delta l, \end{aligned} \tag{2.7}$$

where θ denotes the complementary angle of the angle between the baseline vector ($\mathbf{D}$) and the direction from the earth to the source center ($\mathbf{s}(\Delta l = 0)$). Here, $\frac{D\cos\theta}{\lambda}$ ($\equiv u$) is called as a *'fringe frequency'*. Thus, Δl is represented by using the relation as: $\Delta\phi = 2\pi u\Delta l$.

The radiation spectra of two radiation sources at l_1 and l_2 are denoted as $\hat{E}_1(\nu)$ and $\hat{E}_2(\nu)$, respectively. Then, the intensity distribution on the plane of the sky is represented as $I(\nu, l) = |\hat{E}_1(\nu)|^2\delta(l - l_1) + |\hat{E}_2(\nu)|^2\delta(l - l_2)$. For simplicity, the geometrical delay (τ_g) is assumed to be completely canceled out by its correction (τ_i) for the radiation from the origin ($l = 0$). Then, the phases of the cross power spectra from the radiation source at l_1 and l_2 are represented as:

$$\phi_1 = 2\pi\nu\tau_{g1} = 2\pi u l_1, \quad \phi_2 = 2\pi\nu\tau_{g2} = 2\pi u l_2, \tag{2.8}$$

respectively. Here, τ_{gi} denotes the difference between the geometrical delays for the radiation from the source at the origin and that at l_i. If the radiation from the two sources does not interfere with each other (quasi-monochromatic radiation [1]), their correlation function and cross power spectrum are obtained as the sum of contributions from each radiation source. They are represented as follows:

$$\begin{aligned} C(\tau) &= a_1 a_2^* \int_{-\infty}^{\infty} d\nu \left(|\hat{E}_1(\nu)|^2 \exp(2\pi i\nu(\tau + \tau_{g1})) + |\hat{E}_2(\nu)|^2 \exp(2\pi i\nu(\tau + \tau_{g2}))\right) \\ &= a_1 a_2^* \int_{-\infty}^{\infty} d\nu \left(|\hat{E}_1(\nu)|^2 \exp(2\pi i u l_1) + |\hat{E}_2(\nu)|^2 \exp(2\pi i u l_2)\right) \exp(2\pi i\nu\tau), \end{aligned} \tag{2.9}$$

$$\hat{C}(\nu) = \int_{-\infty}^{\infty} d\tau\, C(\tau) \exp(-2\pi i \nu \tau)$$
$$= a_1 a_2^* \int_{-\infty}^{\infty} d\tau \int_{-\infty}^{\infty} d\nu' \left(|\hat{E}_1(\nu')|^2 \exp(2\pi i \nu' \tau_{g1}) + |\hat{E}_2(\nu')|^2 \exp(2\pi i \nu' \tau_{g2}) \right)$$
$$\times \exp(2\pi i (\nu' - \nu)\tau)$$
$$= a_1 a_2^* \left(|\hat{E}_1(\nu)|^2 \exp(2\pi i u l_1) + |\hat{E}_2(\nu)|^2 \exp(2\pi i u l_2) \right). \quad (2.10)$$

As for the case with more radiation sources, the intensity distribution is represented as: $I(\nu, l) = \sum_k |\hat{E}_k(\nu)|^2 \delta(l - l_k)$. The cross power spectrum is then obtained as: $\hat{C}(\nu) = a_1 a_2^* \sum_k |\hat{E}_k(\nu)|^2 \exp(2\pi i u l_k)$. For a continuous source distribution, the cross power spectrum is represented by an integral as:

$$\hat{C}(\nu) = a_1 a_2^* \int_{\text{source}} dl\, |\hat{E}(\nu, l)|^2 \exp(2\pi i u l)$$
$$= a_1 a_2^* \int_{\text{source}} dl\, I(\nu, l) \exp(2\pi i u l). \quad (2.11)$$

Here, $\hat{C}(\nu)$ can be regarded as a function of the fringe frequency (u), which is denoted as $V(\nu, u)$. $V(\nu, u)$ is called as a *'complex visibility'*. When the intensity distribution $I(\nu, l)$ equals 0 outside the radiation source, it is represented as

$$V(\nu, u) = a_1 a_2^* \int_{-\infty}^{\infty} dl\, I(\nu, l) \exp(2\pi i u l) \quad (2.12)$$

by extending the integral area to all l. Here, $V(\nu, u)$ is the Fourier transform function of $I(\nu, l)$ with respect to u and l. Hence, once the visibility data $V(\nu, u)$ at a specific frequency (ν) is obtained by observations for every fringe frequency (u), the intensity distribution of the radiation source $I(\nu, l)$ at the frequency (ν) is derived from the visibility data by applying Fourier transformation to it. An interferometer is a device to observe $V\left(\nu, u = \frac{D\cos\theta}{\lambda}\right)$ for every baseline vector $\mathbf{D}$ of each pair of two antennas.

2.2 ALMA

Atacama Large Millimeter/submillimeter Array (ALMA) (Fig. 2.3) is the largest radio interferometer working in the millimeter- and submillimeter-wave regions, which is located at the high altitude (5000 m) site (Llano de Chajnantor) of the Atacama desert of northern Chile. It was constructed by the trilateral international partnership among East Asia (including Japan), North America, and Europe. The construction was started in 2002, and the early science operation was started with the limited number of the array antennas in September 2011. Finally, the inauguration

Fig. 2.3 Atacama Large Millimeter/submillimeter Array (ALMA)

of ALMA was declared in March 2012. ALMA is now being operated by the Joint ALMA observatory, which is funded by ESO, AUI/NRAO, and NAOJ.

2.2.1 ALMA Site

Llano de Chajnantor is an excellent site for millimeter and submillimeter interferometry. This site is very dry over the whole year. This is essential, because the submillimter-wave are strongly absorbed by the water vapor contained in the atmosphere. At this site, the precipitable water vapor (pwv) is often below 0.5 mm, which allows us to conduct good observations even at the frequency from 400 GHz to 1000 GHz. Moreover, the site is reasonably flat over a large area, and hence, the large array of antennas can be extended with the maximum baseline length of 18.5 km. With such a long-baseline configuration, an angular resolution as high as $0.''01$ is achieved at the wavelength of 1 mm (300 GHz). This angular resolution is higher by about ten times than that of Subaru Telescope of $0.''2$ ($\lambda = 2.15$ μm).

2.2.2 Antennas

ALMA consists of 54 antennas with a diameter of 12 m and 12 antennas with a diameter of 7 m. Thus the total number of antennas is currently 66. The 12 m and 7 m antennas have a surface accuracy better than 25 and 20 μm [3], which is much smaller than the wavelength of the highest frequency. The highly accurate surface ensures a high aperture efficiency. Fifty 12-m antennas compose the main array (or the 12 m array). The main array is used for observations at various angular resolution. The array configuration is changed little by little (almost monthly) by moving some antennas from the stations to the others. Thus the configuration is extended from the most compact configuration to the most extended configuration for about a half year, and then back to the most compact configuration for another half year. This allows the observers to choose the best angular resolution needed their observations. Four 12-m antennas (the total power array) and twelve 7-m antennas constitute Atacama Compact Array (ACA) or Morita Array. Since the baseline length between neighboring antennas can be relatively short (7 m in principle) in comparison with that between the 12 m antennas (12 m in principle), the low spatial frequency component can be recovered by the ACA system. Furthermore, the total power array provides the visibility data for the zero baseline spacing.

2.2.3 Receivers

ALMA covers all the atmospheric windows from 30 GHz to 950 GHz. For this purpose, 10 receiver bands (Band 1 to 10) are defined as listed in Table 2.1, and a high-sensitivity receiver is prepared for each receiver band. The receiver bands cover the atmospheric transmission windows [3]. At this moment (in 2017), Band 1 and Band 2 have not been commissioned yet, and Band 5 is now being installed. All the receivers except for Band 1 make use of the superconducting mixers called SIS (superconductor-insulator-superconductor) mixers whose receiver temperature is a few times the quantum noise ($h\nu/k$). All the receivers equip two mixer systems sensitive to the orthogonal linear polarization signals. Moreover, the 2 sideband (2SB) mixer is employed except for the higher frequency bands (Band 9 and 10), where

Table 2.1 Frequency Ranges of ALMA Bands[a]

ALMA Band	Frequency Range (GHz)
1	31–45
2	67–90
3	84–116
4	125–163
5	162–211
6	211–275
7	275–373
8	385–500
9	602–720
10	787–950

[a]Taken from ALMA website (http://www.almaobservatory.org/en/about-alma/how-does-alma-work/technology/front-end)

the upper sideband (USB) and the lower sideband (LSB) signals are separated in the mixer and are observed simultaneously. These state-of-the art technologies realizes very high sensitivity of ALMA in combination with the good observing condition of the site. Roughly speaking, the sensitivity for point sources are higher than those of the other single-dish telescopes and interferometers by two orders of magnitude.

2.2.4 Backends

ALMA has a strong bank of backend correlators. The correlators for the main array can be used in the various modes with different frequency resolutions and total bandwidths. They are summarized in Table 2.2. For the continuum observation or the observation of broad spectral lines (e.g. external galaxies), the mode with a lower spectral resolution and a wide bandwidth is useful to maximize the sensitivity. On the other hand, the high resolution mode is employed to study the kinematic structure of the source in detail. It is also useful for line identification in a congested spectrum. The correlator setup is quite flexible. One can set two sub-bands each in USB and LSB, and four chunks of the correlator each in sub-bands. Hence, 16 frequency ranges can be observed simultaneously in total. This is very powerful for astrochemical studies, because various molecular lines can be observed at once under the identical observation condition.

The high frequency resolution is also powerful for astrophysical studies, because it allow us to know the velocity of the molecular gas. If the molecule is moving at a certain velocity, the line frequency of the molecule shifts from its rest frequency by the Doppler effect. The doppler shifted frequency (f') is represented as:

$$f' = \frac{c}{c+v} f, \tag{2.13}$$

Table 2.2 Frequency Resolutions and Total Bandwidths of ALMA Correlators[a]

Total bandwidth (MHz)	Number of channels	Frequency resolution (kHz)	Velocity resolution at 100 GHz (km s^{-1})
2000	128	15625	46.84
1875	3840	488	1.46
936	3840	244	0.73
469	3840	122	0.37
234	3840	61	0.18
117	3840	30.5	0.09
58.6	3840	15.3	0.05

[a]The total bandwidths, number of channels, and frequency resolutions are taken from [3]

where c denotes the speed of light, v the velocity of the molecule along the line of sight, and f the rest frequency of the molecule. In this study, v is sufficiently smaller than c, and thus f' can be approximated by $\left(1 - \frac{v}{c}\right) f$. The ALMA correlator can resolve even the fine change in f due to the Doppler effect. The velocity resolution with the rest frequency f of 100 GHz is also summarized in Table 2.2. The velocity resolution better than the sound speed of the molecular-cloud gas (~0.2 km s^{-1}) can readily be achieved.

2.3 Observations with Interferometers

2.3.1 Calibration

The output data from an interferometer ('*observed complex visibility function*') suffer from amplitude variations and phase delays due to the array hardware as well as the atmospheric absorption and scintillation. Thus, calibrations are required to obtain the true visibility function, which is physically related to the angular and spectral distribution of the target source, from the raw observed data. Here, the major calibrations required for interferometers are briefly described.

2.3.1.1 Phase Calibration

The phase calibration is the calibration for the phase of the visibility function. The phase of the visibility function, which contains the information of the position of the radiation source distribution, suffers from the atmosphere delay over each antenna and the instrumental delay in the array. In order to eliminate these effects, the phase calibration is performed toward a well-known calibrator source every several minutes (e.g. 8 minutes in the ALMA observations in Chap. 6) during the observation. A phase calibrator source should be close to the science target, and its position, spectrum, and flux density should accurately be known. In order to determine the phase center accurately, the angular size of the phase calibrator has to be sufficiently small in comparison with the beam size of the observation. Thus quasars are often used as the phase calibrator because their true visibility function can be modeled. The phase calibrator sources used in this study are listed in Table 2.3. If the target source is enough bright, the phase calibration can be done with the target source itself. This procedure is called as '*self-calibration*'. The self-calibration is applied to the analyses for IRAS 16293−2422 Source B and L483 (Chaps. 7 and 8).

In addition, a water vapor radiometer (WVR) is used for the short-term fluctuation of the phase delay due to the atmosphere. Since the delay is caused by the water vapor, it is related to the precipitable water vapor (pwv). The delay can be corrected by measuring the temporal fluctuation of the pwv by WVR equipped on each antenna.

This method, which is automatically done in the observation, allows us to achieve the high angular-resolution performance of ALMA.

2.3.1.2 Bandpass Calibration

The frequency response of the interferometer is corrected by the bandpass calibration. The bandpass calibration is performed with a bright calibrator source, which has a known radio spectral index without spectral features. Some bright quasars are typically used as a bandpass calibrator. The frequency response is almost stable over many hours, and nearly independent of the sky position of the radio source. Thus, the bandpass calibration is performed only the start or end of the observation, or additionally a few times during the observation. The bandpass calibrations typically takes a half or one hour. However, if the observation for other calibrations is performed toward a sufficiently bright source, it can be also used for the bandpass calibration. The bandpass calibrator sources used in this study are listed in Table 2.3.

2.3.1.3 Gain and Flux Density Calibrations

The absolute flux density of the target source is obtained by applying the gain calibration and the flux density calibration. The relative amplitude between the phase calibrator and the target source is determined by the gain calibration. The scale from the observed amplitude to the absolute flux density is usually obtained by observations toward a Solar System object. The flux density of the flux calibrator is monitored by the ALMA observatory and its recent value is accurately known. The observations toward flux calibrator should be performed in each phase calibration. However, if the antenna gains are sufficiently stable, the observation for the flux calibration is performed only once at the start of the observation in order to scale the absolute flux of the phase calibrator. Then the absolute flux of the target source can be derived from the relative amplitude to the phase calibrator. For the ALMA Cycle 5 observation, the flux density model accuracy is 3–15% at Band 3-9 for the Solar System objects. The flux calibrator sources used in this study are listed in Table 2.3.

Table 2.3 Calibrator Sources Used in This Study

Chapter	Science target	Phase calibrator	Bandpass calibrator	Flux calibrator
Chapter 4	L1527	J0510+180	J0423−013	Callisto
Chapter 5	IRAS 15398−3359	J1517−243	J1256−057, J1924−292	Mars, Titan
Chapters 6 and 7	IRAS 16293−2422	J1626−2951, J1700−2610, J1625−2527	J1733−1304, J1700−2610, J1517−2422, J1427−4206	Titan
Chapter 8	L483	J1733−1304	J1733−1304	Titan

2.3.2 Observed Intensity Distribution

In the actual observational studies, the visibility data (Eq. 2.12) is two-dimensional, and is represented as $V(\nu; u, v)$. Here, the fringe frequencies u and v are represented as:

$$u = \frac{D_x \cos\theta}{\lambda}, \quad v = \frac{D_y \cos\phi}{\lambda}, \tag{2.14}$$

where θ and ϕ are the complementary angles of the apparent angle between the baseline vector (**D**) and the direction from the earth to the source center (**s**) projected onto the (l, n)- and (m, n)-plane, respectively. D_x and D_y denote the component of **D** along the x- and y-axis. The intensity distribution of the radiation source on the plane of the sky $I(\nu; l, m)$ is obtained by applying two dimensional Fourier transformation to $V(\nu; u, v)$ for u and v. Practically, the number of antennas of an interferometer is limited, so that $V(\nu; u, v)$ is observed only for a limited number of the (u, v) values, and the data for the remaining (u, v) values are missing. Hence, the observed data $\tilde{V}(\nu; u, v)$ can be regarded as the product of the real visibility data $V(\nu; u, v)$ and a two dimensional function $U(u, v)$. $U(u, v)$ is called as the '*(u, v)-coverage*'. It has a non-zero value for (u, v) corresponding to a certain baseline vector in an interferometer, while it equals 0 for the other (u, v) without the observed visibility data. Since the array of the antennas rotates relative to the target source due to the earth rotation, each pair of two antennas draws a part of a conic section in the (u, v)-plane. This effect can be used for a better coverage of the (u, v)-plane.

The image of the intensity distribution $\tilde{I}(\nu; l, m)$ is obtained by Fourier transformation of the observed visibility data $\tilde{V}(\nu; u, v)$. $\tilde{I}(\nu; l, m)$ does not reproduce the real intensity distribution $I(\nu; l, m)$ perfectly, as shown in Fig. 2.4, due to an imperfect (u, v)-coverage of the visibility data. The fringe frequency (u, v) is related to the spatial frequency component in $I(\nu; l, m)$. The values of $V(\nu; u, v)$ for a set of small values of (u, v) correspond to the low spatial frequency component of $I(\nu; l, m)$, or an extended component, while those for a set of large values of (u, v) do the high frequency component, or a compact component. Hence, the integrated value of the observed intensity $\left(\int dldm\, \tilde{I}(\nu; l, m)\right)$ is generally lower than that of

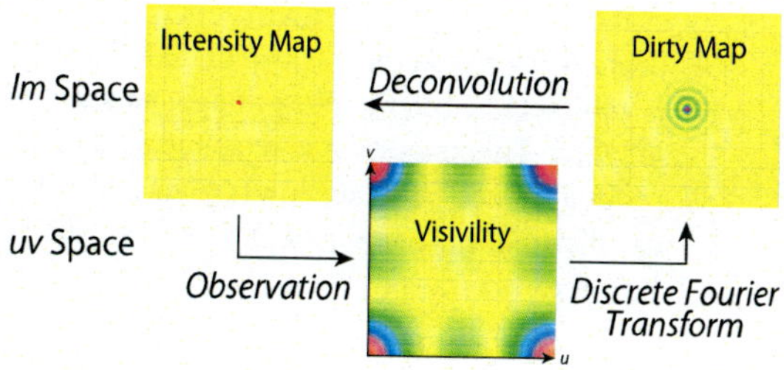

Fig. 2.4 Relations between the maps of the real intensity distribution, the observed visibility data, and the observed intensity distribution

the real intensity $\left(\int dldm\ I(\nu; l, m)\right)$, and their difference is called as *'missing flux'*. Specifically, if the visibility data with (u, v) close to the origin is missing, extended components of the radiation source is not reproduced. It is called as *'resolving out'*. The Fourier-transformed function of $U(u, v)$ is defined as $B(l, m)$. Since $\tilde{I}(\nu; l, m)$ is the Fourier-transformed function of the product of $V(\nu; u, v)$ and $U(u, v)$, it is represented as the convolution of $I(\nu; l, m)$ with $B(l, m)$:

$$
\begin{aligned}
\tilde{I}(\nu; l, m) &= \int dudv\ V(\nu; u, v) \cdot U(u, v) \exp(-2\pi i(ul + vm)) \\
&= I(\nu; l, m) * *B(l, m).
\end{aligned}
\tag{2.15}
$$

Here, $B(l, m)$ is called as *'synthesized beam'*. If $U(u, v)$ is unity for all (u, v) values, $B(l, m)$ is represented as $\delta(l)\delta(m)$. In reality, $U(u, v)$ takes non-zero values only for the observed (u, v) values, and $B(l, m)$ has a width inversely proportional to the maximum u and v values. If the visibility data with the large (u, v) values are missing, compact components of the radiation source are not reproduced. Thus, the longest baseline (the largest (u, v) value) determines the angular resolution of the interferometer.

2.3.3 CLEAN Method

$\tilde{I}(\nu; l, m)$ obtained with using the Eq. (2.15) is called as *'dirty map'*. It is the convolution function of the real intensity distribution $I(\nu; l, m)$ with the synthesized beam $B(l, m)$ as shown in Fig. 2.4. A dirty map suffers from a spurious wavy distributions on the real intensity distribution, which is called as a *'side lobe'*. In order to reproduce the image of the observed intensity distribution closer to the real one, a *'deconvolution'* procedure, which is an operation to eliminate the effect of the synthesized beam from the image, is usually performed in the analysis of interferometer data. Since the real $V(\nu; u, v)$ for missing (u, v) is unknown, the image obtained by the deconvolution is not unique due to this uncertainty. In this study, the *'CLEAN'* method is employed to obtain an appropriate image.

CLEAN is a method of the deconvolution suggested by [2], and is commonly used in radio interferometry. In the CLEAN method, the intensity distribution of the radiation source is modeled by a set of point sources. A point source is supposed at the intensity peak in a dirty map, and its convolved image with the synthesized beam (or *'dirty beam'*) is calculated. The convolved image, including its side lobe, is subtracted from the dirty image, and thus the effect of the spurious component is reduced. This operation is repeated specified times or until the peak intensity in the residual image becomes lower than a specified value. After that, the residual image has distributions, side lobe, and noise, which are expected to be much weaker than the modeled image. Then, the modeled image, which is a set of point sources, is convolved with an ideal beam function (or *'CLEAN beam'*), and is added together

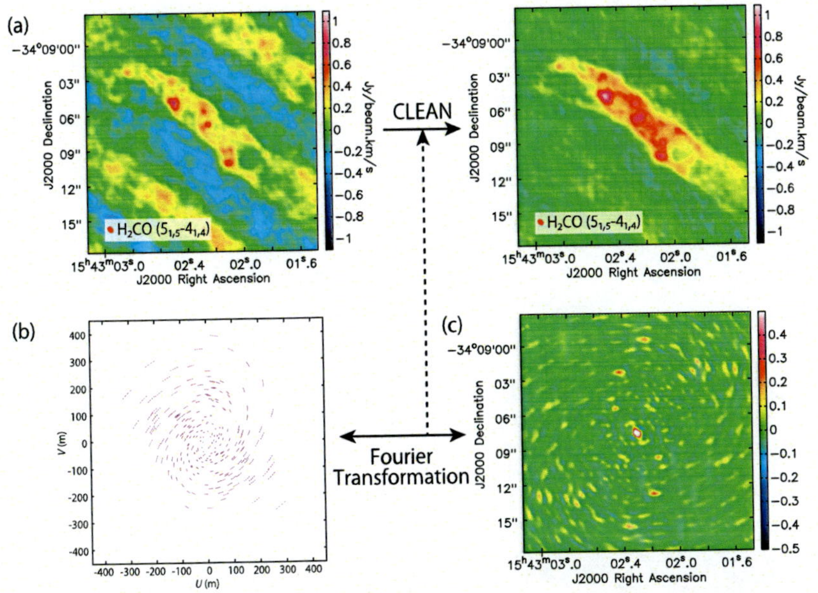

Fig. 2.5 **a** Dirty map and the CLEANed image of the H_2CO $(5_{1,5} - 4_{1,4})$ line toward IRAS 15398−3359 observed with ALMA. **b** (u, v)-coverage of this observation. **c** Point spread function of this observation, which corresponds to the dirty map observed toward a ideal point source with the intensity of unity. These panels show the case for the frequency (~351.763 GHz) almost correspond to the doppler-shifted frequency by the systemic velocity of the source (~5 km s^{-1})

with the residual image to estimate the real intensity distribution $I(\nu; l, m)$. In the process, the Gaussian beam with the same FWHM (full width at half maximum) as that of the synthesized beam $B(l, m)$ is commonly employed as the CLEAN beam. Figure 2.5a shows examples of the dirty map and the image obtained after applying the CLEAN method. These are the distribution of the H_2CO $(5_{1,5} - 4_{1,4})$ emission observed toward the low-mass Class 0 protostellar source IRAS 15398−3359 with ALMA (Chap. 5). Twenty-five antennas were used in these observations. These images are derived from the observed visibility data with a (u, v) coverage shown in Fig. 2.5b. Figure 2.5c is the *'point spread function'*, which is the convolved image of a point source with the dirty beam.

References

1. Born M, Wolf E (1959) Principles of optics
2. Högbom JA (1974) Aperture synthesis with a non-regular distribution of interferometer baselines. A&AS 15:417
3. ALMA Partnership (2017). ALMA Cycle 5 Technical Handbook

Chapter 3
Model Calculation

3.1 Introduction

As shown in Chaps. 4–8, the disk and envelope components associated to the protostar and the bipolar outflow launched from the vicinity of the protostar are detected. These components are heavily contaminated with one another, and thus their disentanglement is essential to investigate the geometrical and kinematic structure around the protostar. For this purpose, models of an envelope, a disk, and an outflow are employed to calculate their kinematic structure to disentangle the contaminated components obtained in the observations. Moreover, the models are useful to evaluate physical parameters, such as the protostellar mass, the specific angular momentum of the gas, and the inclination angle.

It is generally thought that the mid-plane of the disk/envelope system and the outflow axis are almost perpendicular to each other. Their configuration thus can be constrained if one of them is characterized. In Chaps. 4–8, the characterization is started with the component detected more clearly in each source. For instance, the disk/envelope system is more clearly delineated than the outflow component in L1527 (Chap. 4), while the outflow is more clearly delineated in IRAS 15398−3359 (Chap. 5).

3.2 Infalling-Rotating Envelope Model

3.2.1 Configuration of the Infalling-Rotating Envelope Model

A ballistic model is constructed to investigate infalling-rotating envelopes [7]. The basic concept of this model is introduced to explain the kinematic structure of the infalling-rotating envelope of L1527 [11]. Here, it is improved to consider the velocity

Y. Oya, *A Few Tens au Scale Physical and Chemical Structures Around Young Low-Mass Protostars*, Springer Theses, https://doi.org/10.1007/978-981-19-1708-0_3

field in the three dimensional space with the convolution by the beam size and the velocity resolution.

Figure 3.1 shows the schematic illustration of the 'infalling-rotating envelope' model. In this model, the gas is simply assumed to be falling and rotating under the gravity of the central protostar. The motion of the gas is approximated by the particle motion, ignoring effects of gas pressure, magnetic field, self gravity, and so on. Because of the energy and angular momentum conservation, the gas cannot fall inward of a certain radius, or the 'perihelion'. This position is called as 'centrifugal barrier'. The radius of the centrifugal barrier (r_{CB}) is represented as:

$$r_{\mathrm{CB}} = \frac{j^2}{2GM}, \tag{3.1}$$

where G is the gravitational constant, M is the protostellar mass, and j is the specific angular momentum of the gas. It is the radius at which all the kinetic energy is converted to the rotational energy. It is a half of the centrifugal radius (r_{CR}), where the gravitational force and the centrifugal force balance each other out:

$$\begin{aligned} r_{\mathrm{CR}} &= \frac{j^2}{GM} \\ &= 2r_{\mathrm{CB}}. \end{aligned} \tag{3.2}$$

The rotation and infall velocities (v_{rot} and v_{fall}) of the gas at the distance of r to the protostar are represented as follows:

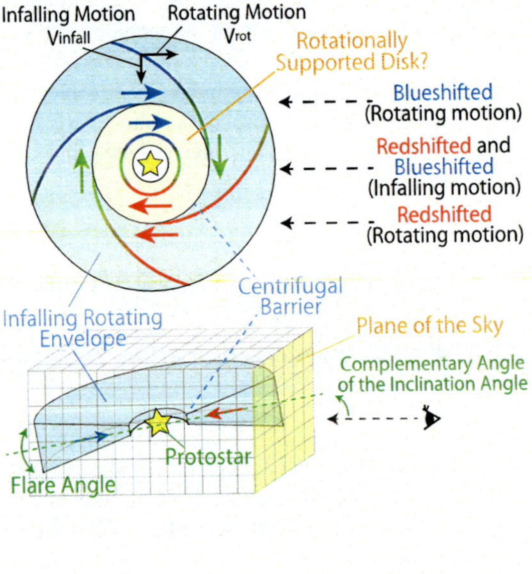

Fig. 3.1 Schematic illustration of the infalling-rotating envelope model. The space and the velocity field are sectioned into meshes

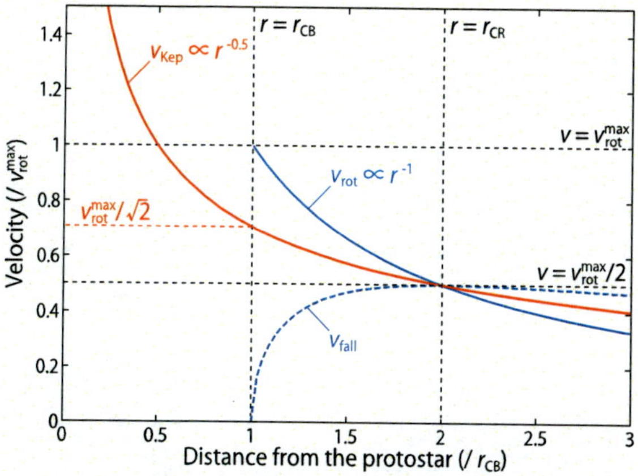

Fig. 3.2 The velocity of the gas in the infalling-rotating envelope and Keplerian disk model as a function of the distance from the protostar. The horizontal axis represents the distance from the protostar (r) normalized by the radius of the centrifugal barrier (r_{CB}), and the vertical axis represents the velocity (v) normalized by the rotation velocity of the infalling-rotating envelope model at the centrifugal barrier $\left(v_{rot}^{max}\right)$. The blue solid and dashed lines represent the rotation (v_{rot}) and infall (v_{fall}) velocities of the infalling-rotating envelope model. v_{fall} equals 0 at the centrifugal barrier ($r = r_{CB}$), and v_{rot} takes its maximum value there. On the other hand, v_{fall} takes the maximum value at the centrifugal radius ($r_{CR} = 2r_{CB}$). The red solid line represents the Keplerian velocity (v_{Kep}). All of v_{rot}, v_{fall}, and v_{Kep} take the same value $\left(v = v_{rot}^{max}/2\right)$ at the centrifugal radius (r_{CR})

$$\begin{aligned} v_{rot} &= \frac{j}{r} \\ &= \frac{1}{r}\sqrt{2GMr_{CB}}, \end{aligned} \tag{3.3}$$

$$\begin{aligned} v_{fall} &= \sqrt{\frac{2GM}{r} - v_{rot}^2} \\ &= \frac{1}{r}\sqrt{2GM\left(r - r_{CB}\right)}. \end{aligned} \tag{3.4}$$

Thus, the velocity field is determined by M and r_{CB}. The inclination angle (i) of the disk/envelope system also affects the apparent velocity along the line of sight. At the centrifugal barrier, v_{fall} equals to 0, and v_{rot} takes its maximum value. On the other hand, v_{fall} takes its maximum value at the centrifugal radius. The values of v_{rot} and v_{fall} as a function of the radius from the protostar are plotted in Fig. 3.2.

In this model, the distribution of the gas is assumed to have a power-law. The power-law of $r^{-1.5}$ corresponds to the density profile of an infalling cloud (e.g. [3, 6, 13]). An optically thin condition is also assumed, where the intensity of the line emission is proportional to the column density along the line of sight. Namely, excitation effects and radiative transfer effects are not considered. These assumptions are rather arbitrary in this study, because the main interest in the model analysis is on the velocity field of the gas around the protostar.

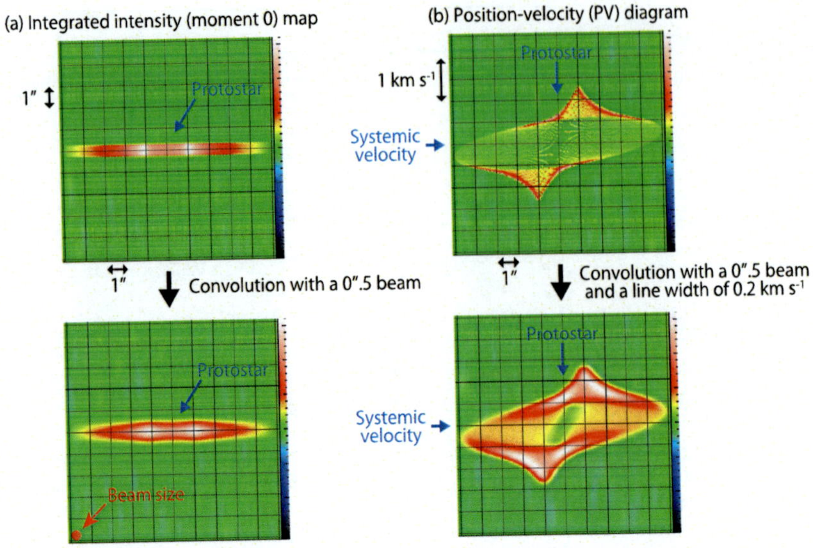

Fig. 3.3 Effects of the convolution of the line emission with an intrinsic Gaussian profile (FWHM = 0.2 km s^{-1}) and a Gaussian beam (0.″5 × 0.″5) in the infalling-rotating envelope model. The color scales represent the intensity of the line emission. Panels **a** and **b** show the integrated intensity map and the position velocity (PV) diagram along the disk/envelope system, respectively. In these models, the inclination angle (i) of 90° (an edge-on configuration) is employed for simplicity

The spectral line is assumed to have an intrinsic Gaussian profile with a certain line width, and the emission is convolved with a Gaussian beam with a certain full width at half maximum (FWHM) (Fig. 3.3). The intrinsic line width and the beam size are employed depending on each source. The mesh sizes are also appropriately chosen for each source.

Figure 3.4 shows an example of the results of the infalling-rotating envelope model. Its physical parameters are summarized in the caption. Figure 3.4a shows the integrated intensity map. The envelope is assumed to have an edge-on configuration extended along the east-west axis, where the left- and right-hand sides correspond to east and west, respectively. The integrated intensity relative to its peak value in the panel is shown in a color scale. The protostar with a mass (M) of 0.1 $M_\odot$ exists at the central position in Fig. 3.4a. The distance to the source from the Sun (d) is set to be 100 pc, where 1″ corresponds to 100 au. The envelope has an outer radius (R) of 500 au, outside which there is no molecule. The radius of the centrifugal barrier is 100 au, and the molecular density is zero inside the centrifugal barrier. The integrated intensity is the highest around the centrifugal barrier.

Figure 3.4b shows the position-velocity (PV) diagram along the blue arrow shown in Fig. 3.4a. The angular offset of 0″ corresponds to the protostellar position. The vertical axis represents the line-of-sight velocity of the molecules relative to the systemic velocity of the source. In Fig. 3.4b, a spin-up feature can be seen along the

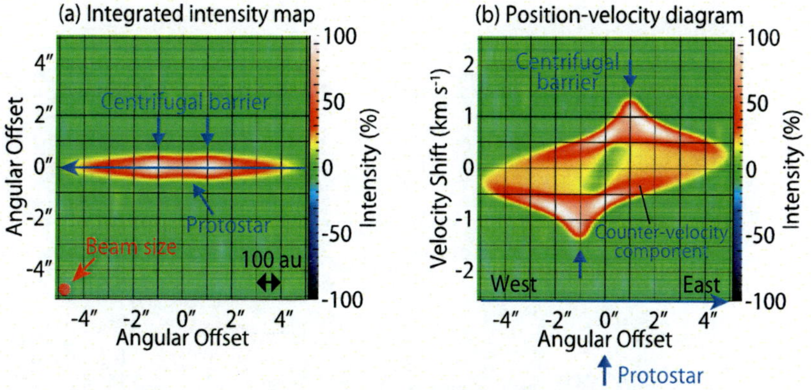

Fig. 3.4 The results of the infalling-rotating envelope model. Panels **a** and **b** show its integrated intensity map and position-velocity diagram, respectively. The position axis in panel (**b**) is taken along the blue horizontal arrow shown in panel (**a**), along which the mid-plane of the envelope is extended. The physical parameters of the model is as follows; the distance to the source from the Sun (d) of 100 pc, the protostellar mass (M) of 0.1 $M_{\odot}$, the radius of the centrifugal barrier ($r_{\rm CB}$) of 100 au, the inclination angle of the disk/envelope system (i) of 90° (0° for a face-on configuration), and the outer radius of the envelope (R) of 500 au. The scale height of the envelope is assumed to be 50 au independent of the radius. The mesh size for the calculation is 0.″08 (8 au), and 128 × 128 × 128 meshes are applied to cover the cubic space of 10.″24 × 10.″24 × 10.″24 (1024 × 1024 × 1024 au^3). The mesh size for the velocity axis is 0.02 km s^{-1}, and 256 meshes are applied to cover the velocity range from −2.56 km s^{-1} to 2.56 km s^{-1}. The intensity is convolved with the intrinsic line width of 0.2 km s^{-1} and the beam of 0.″5×0.″5

east-west axis; the rotation velocity of the molecules increases as approaching to the protostar till the centrifugal barrier. The velocity takes its maximum and minimum values around the centrifugal barrier, where the velocity is positive (red-shifted) and negative (blue-shifted) in the eastern and western sides of the protostar, respectively. Toward the protostellar position, only a velocity shift due to the infall motion can be seen. The infall motion can also be confirmed as the counter velocity component (Fig. 3.4b).

3.2.2 *Infalling-Rotating Envelope Model with Various Physical Parameters*

In this study, the physical parameters of low-mass protostellar sources are evaluated by comparing the observed kinematic structure of the gas and the infalling-rotating envelope model. For this purpose, it is demonstrated how sensitive to the physical parameters the model results are for some cases.

The infalling-rotating envelope model has three major physical parameters; M, $r_{\rm CB}$, and i. The apparent distribution of the line emission projected onto the plane of the sky is determined by $r_{\rm CB}$ and i. Figure 3.5 shows the integrated intensity

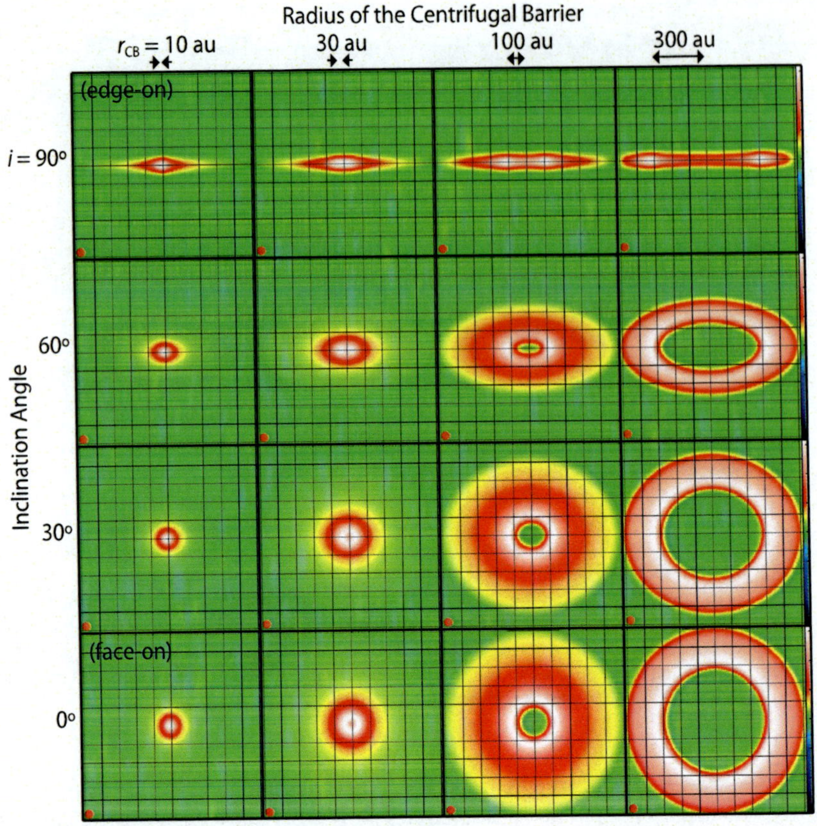

Fig. 3.5 Integrated intensity (moment 0) maps of the infalling-rotating envelope model with various sets of the radius of the centrifugal barrier ($r_{\rm CB}$) and the inclination angle of the disk/envelope system (i). The other physical parameters for the models are set as follows; the distance from the Sun d = 100 pc, the protostellar mass $M = 0.1\ M_\odot$, and the outer radius of the envelope R = 500 au. The FWHM values of the intrinsic line width and the Gaussian beam are set to be 0.2 km s^{-1} and 0.″5, respectively. The uniform scale height of the envelope of 50 au is assumed. The mesh sizes are set to be 0.″08 and 0.01 km s^{-1}

maps for various sets of $r_{\rm CB}$ and i. Here, i of 0° and 90° corresponds to face-on and edge-on configurations, respectively. The FWHM value of the Gaussian beam (angular resolution) is 0.″5 (50 au). The other physical parameters are summarized in the caption of Fig. 3.5. With the edge-on configuration (i = 90°), the integrated intensity maps show a flattened feature. The maps with $r_{\rm CB}$ of 10 and 30 au show single-peaked distributions. Their centrifugal barriers seem to be almost unresolved with the beam of 0.″5. On the other hand, the maps with $r_{\rm CB}$ of 100 and 300 au show double-peaked distributions. The peak positions seem to correspond to the positions of their centrifugal barriers; namely, their centrifugal barriers are spatially resolved

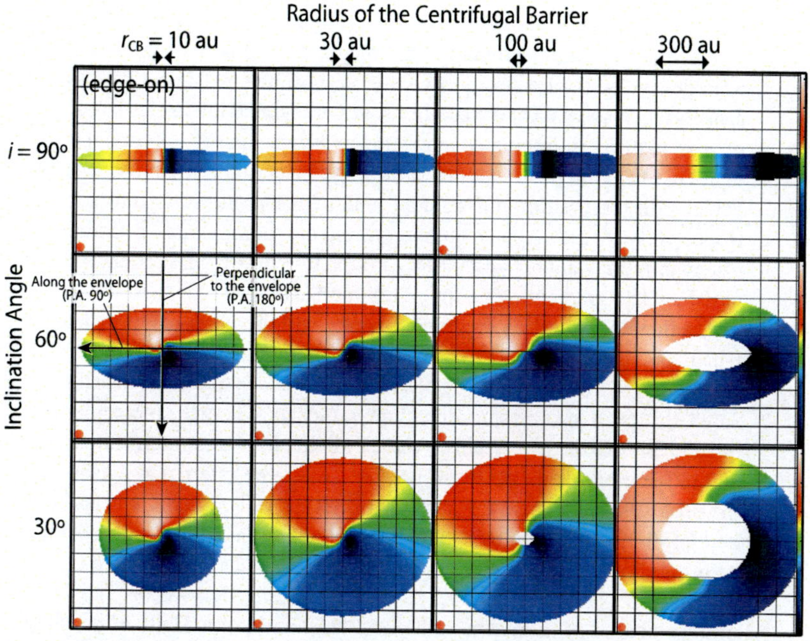

Fig. 3.6 Maps of the velocity field (moment 1 maps) of the infalling-rotating envelope model with various sets of r_{CB} and i. The other physical parameters are set to be as the same as those in Fig. 3.5. The color scale represents the averaged velocity-shift (Eq. 3.5). The color maps are shown only at the positions where the integrated intensity (Fig. 3.5) is larger than 1% of the peak integrated intensity. Thus, the data at the positions with a weaker integrated intensity than this criteria are shown in white. The black arrows in the panel with r_{CB} of 10 au and i of 60° represent the directions along which the PV diagrams in Figs. 3.7, 3.8 and 3.9 are prepared

with the beam. At i of 60° (nearly edge-on), the hole of the distribution is clearly seen in the panels for r_{CB} of 100 and 300 au. Although the model for r_{CB} of 30 au does not show a clear double-peaked distribution at i of 60°, it shows an intensity dip toward the protostar at i of 30° (nearly face-on). With the face-on configuration (i = 0°), the distributions show completely circular or ring-like structures. The intensities have their maximum value around the centrifugal barriers.

The velocity field of the gas is determined by M, r_{CB}, and i. The velocity along the line of sight (v_{shift}) in each mesh is calculated to compare with observations. Figure 3.6 shows the moment 1 maps with the various sets of r_{CB} and i. The FWHM value of the Gaussian beam is 0.″5 (50 au). The other physical parameters are summarized in its caption. The color map represents the average velocity shift weighted by the intensity ($\bar{v}$). It is calculated by using the following equation:

$$\bar{v} = \frac{\int vI(v)dv}{\int I(v)dv}, \tag{3.5}$$

where $I(v)$ denotes the intensity at the velocity v at the position. The denominator in the Eq. (3.5) corresponds to the integrated intensity at the position. Thus, the averaged velocity at a position having an infinitesimal integrated intensity has no sense. Hence, the data at the positions with integrated intensities smaller than the 1% relative to the peak integrated intensity are dropped in Fig. 3.6, and they are shown in white in the moment 1 maps. With the edge-on configuration ($i = 90°$), the averaged velocity in the eastern and western side of the protostar is red- and blue-shifted, respectively. These velocity shifts represent the rotation motion around the protostar. The maximum velocity shift is seen around the centrifugal barrier. The velocity does not show any gradient along the north-south axis.

On the other hand, velocity gradients along the north-south axis are seen in the panels for i of 30° and 60°. These velocity gradients are due to the infall motion. In these models, the southern side of the envelopes face to us, and thus the line emission is red-shifted in the northern side of the protostar, where the molecules are in front of the protostar. In these panels, the velocity fields show skewed features; the most blue- and red-shifted components are seen in the southwestern and northeastern sides of the protostar, respectively, where the projected velocity components of the rotation and infall motions have the same direction along the line of sight. With the face-on configuration ($i = 0°$), the velocity shift is completely symmetric to the mid-plane of the envelope, and thus the average velocity is zero everywhere. The value of v_{shift} is proportional to M as shown in the Eqs. (3.3) and (3.4). Thus, M does not affect the moment 1 map in appearance.

Figure 3.7 shows the position-velocity (PV) diagrams along the arrows shown in Fig. 3.6. The position axes are taken along the direction where the mid-plane of the envelopes are extended (Fig. 3.6). In Fig. 3.7, r_{CB} and i are varied to show how sensitive to these physical parameters the model results are. The FWHM value of the Gaussian beam is 0.″5 (50 au). The other physical parameters are summarized in the caption. With the edge-on configuration ($i = 90°$), the spin-up feature toward the centrifugal barrier is seen with all r_{CB}. The maximum velocity-shift seen at the centrifugal barrier is larger for a smaller r_{CB}. Although the centrifugal barriers are not spatially resolved in the integrated intensity maps for r_{CB} of 10 and 30 au (Fig. 3.5), a velocity gradient is visible between the positions of the centrifugal barriers in their PV diagrams. The infall motion can be confirmed as the counter velocity components. The velocity shifts toward the protostellar position also reflect the infall motion. With i of 30° and 60°, the counter velocity components are not seen for r_{CB} of 100 and 300 au, while they can be seen with r_{CB} of 10 and 30 au. The infall gas in the models with r_{CB} of 100 and 300 au are distant from the protostellar position in the plane of the sky. Therefore, these components are almost outside the beam, and do not have effective contributions in these PV diagrams. With the face-on configuration ($i = 0°$), a velocity gradient due to the rotation motion cannot be seen regardless of r_{CB}. Although the PV diagrams with r_{CB} of 10 au is not smooth like those for larger r_{CB}, this feature does not have any kinematical meaning, but is due to an artificial effect of the insufficient size of the mesh.

Figure 3.8 shows the model results of PV diagrams prepared along various position angles (P.A.) at various i. The P.A.s are taken for every 30°. The P.A.s of '90°' and

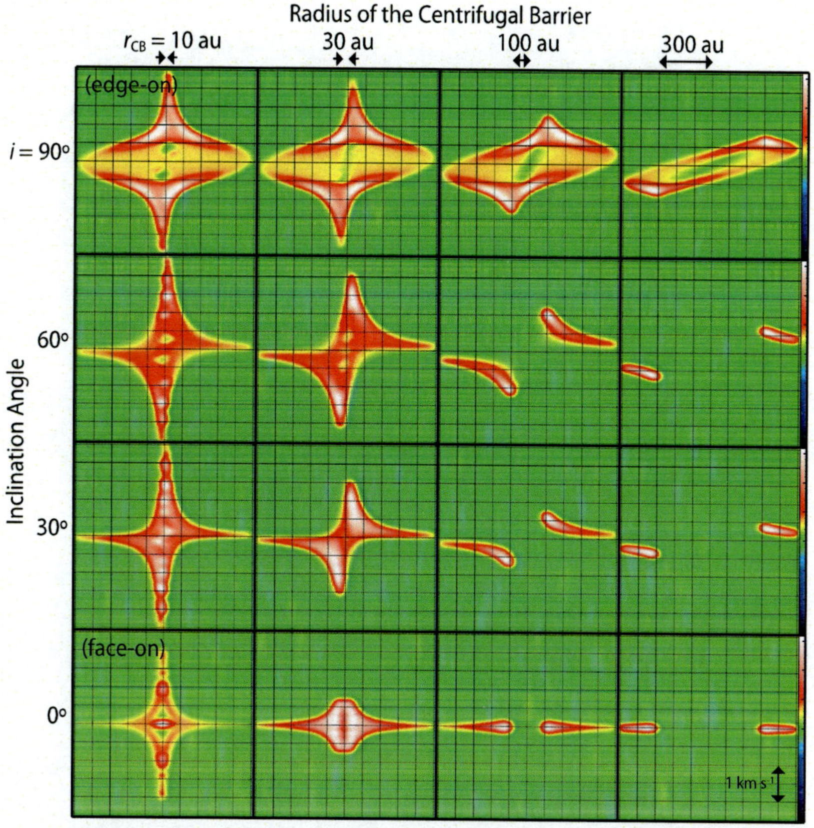

Fig. 3.7 PV diagrams of the infalling-rotating envelope model with various sets of r_{CB} and i. Their position axis is along the arrow (P.A. 90°) shown in Fig. 3.6. It is the direction along which the disk/envelope system is extended. The other physical parameters are set to be as the same as those in Fig. 3.5

'180°' represent the direction along which the mid-plane of the envelope is extended and one perpendicular to it, respectively. The FWHM value of the Gaussian beam is 0.″5 (50 au). The other physical parameters are summarized in the caption. With the edge-on configuration (i = 90°), the distributions look concentrated around the protostar in the PV diagrams with the P.A. of (120° − 240°). A slight velocity gradient can be seen for these P.A.s, except for the P.A. of 180° (the direction perpendicular to the envelope). The velocity shift in the diagram with the P.A. of 180° is due to the infall motion, and it is smaller than that of the rotation motion at the centrifugal barrier. At i of 30° and 60°, the velocity structure changes from P.A. to P.A. The value of the velocity shift is determined by the complex combination of the rotation and infall motions. As shown in Fig. 3.6, the rotation and infall motions cancel each other

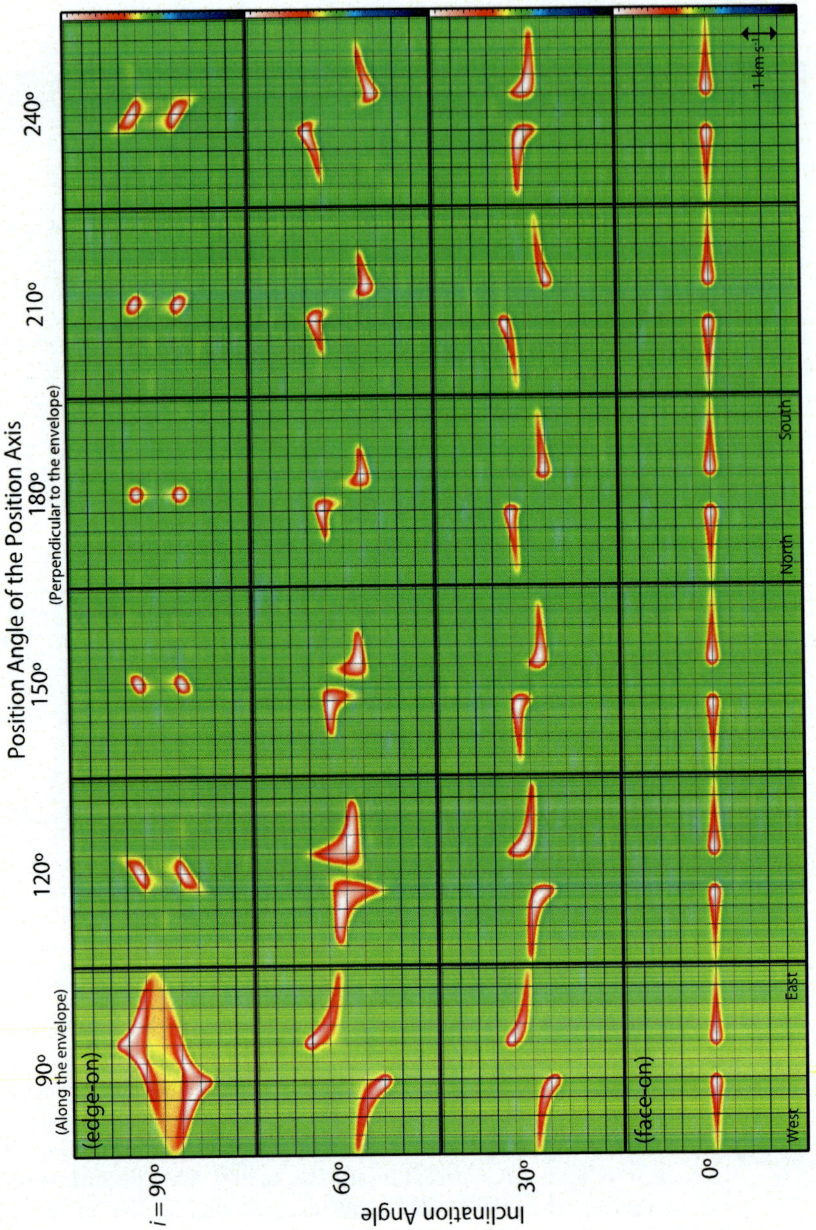

Fig. 3.8 PV diagrams of the infalling-rotating envelope model with various i, where the physical parameters are as follows; $M = 0.1\ M_{\odot}$, $r_{CB} = 100$ au, and $R = 500$ au. The position axes are taken for every 30° from the envelope direction (P.A. 90°)

in the northwestern and southeastern sides of the protostar, while they strengthen each other in the northeastern and southwestern sides. Because of this, the absolute values of the velocity shift tend to be higher in the diagrams with the P.A. of 210° and 240°, where the position axis is along the northeast-southwest direction, than those with the P.A. of 120° and 150°, where the position axis is along the northwest-southeast direction. With the face-on configuration (i of 0°), no velocity gradient can be seen regardless of the P.A.

In Fig. 3.9, the effects of the other physical parameters, such as the protostellar mass (M), the outer radius of the envelope (R), and its scale height, are examined. Although the absolute velocity-shift is larger for larger M, the essential feature of the PV diagrams is not largely affected by M. With a smaller R, the spin-up feature and the counter velocity components are less clear, although the velocity gradient between the centrifugal barriers can be seen. The scale height of the envelope (H)

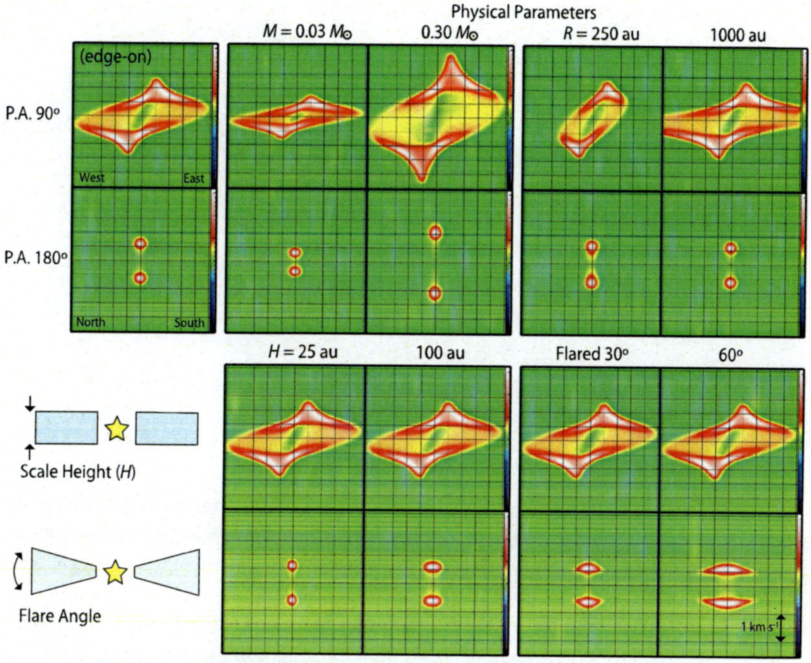

Fig. 3.9 PV diagrams of the infalling-rotating envelope model, whose position axes are along the arrows (P.A. 90° and 180°) shown in Fig. 3.6. The left upper panels show the results of the fiducial model with the following physical parameters; $M = 0.1\ M_\odot$, $r_{CB} = 100$ au, $i = 90°$ (edge-on), $R =$ 500 au, and H of 50 au independent of the radius. In the other panels, one of the following physical parameters is changed; the protostellar mass M (0.03 and 0.3 $M_\odot$), the outer radius of the envelope R (250 and 1000 au), and the scale height of the envelope H (25 and 100 au). The scale height is assumed to be independent of the radius, or assumed to increase as the radius ('flared') with the flare angle of 30° and 60°

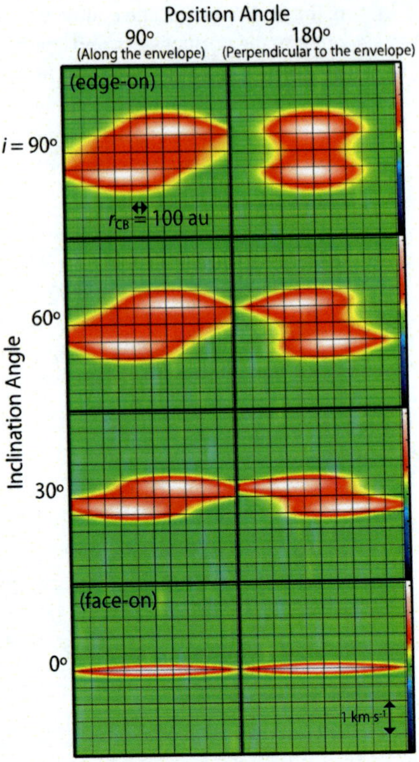

Fig. 3.10 PV diagrams of the infalling-rotating envelope model with various set of i and the position angle, where the other physical parameters are as follows; $M = 0.1\ M_{\odot}$, $r_{CB} = 100$ au, $R = 500$ au, and $H = 50$ au. The position angles of the line along which the position-velocity diagram is prepared are along the mid-plane of the envelope and perpendicular to it in the left and right panels, respectively. The intensity is convolved with the Gaussian beam of $5'' \times 5''$

does not seriously affect the appearance of the PV diagrams with a P.A. of 90° (along the mid-plane extension). With the edge-on configuration, the scale height affects the distribution along the direction perpendicular to the mid-plane extension (P.A. 180°).

Figure 3.10 shows how the appearance of the model results changes with a large beam size (5″). The centrifugal barrier ($r_{CB} = 100$ au) is not spatially resolved enough with the beam (500 au). The velocity gradient due to the rotation motion is recognized for the cases at i of (30° − 90°). Although the counter velocity components cannot be seen with i of 30° and 60° with a beam of 0.″5×0.″5 in Fig. 3.7, they are seen in Fig. 3.10. Since the beam is larger than the apparent distance between the centrifugal barriers in these models, the emission of the infalling gas contributes to the PV diagrams. A similar effect due to the large beam is also seen in the PV diagrams along the direction perpendicular to the envelope; the rotating gas at the centrifugal barrier contributes to these diagrams, and hence, the maximum velocity-shift is as high as the maximum rotation velocity at the centrifugal barrier. In these diagrams, disentanglement of the rotation and infall motion is difficult. In the observational studies, it also makes difficult to find the position angle of the extension of the envelope component. Higher angular-resolution observations are always essential.

3.3 Keplerian Model

In Chaps. 6 and 8, the Keplerian disk model is employed to simulate the disk component. The velocity of the gas at the radius r from the protostar are represented as:

$$v_{\rm Kep} = \sqrt{\frac{GM}{r}}, \quad v_{\rm fall} = 0. \tag{3.6}$$

As shown in Fig. 3.2, the Keplerian velocity takes the lower value than $v_{\rm rot}$ by a factor of $\sqrt{2}$ in the above infalling-rotating envelope model at the centrifugal barrier ($r = r_{\rm CB}$), while it equals to $v_{\rm rot}$ in the infalling-rotating envelope model at the centrifugal radius ($r = 2r_{\rm CB}$).

Figure 3.11 shows the integrated intensity map (moment 0 map), velocity field (moment 1 map), and PV diagrams of the Keplerian model. In these models, the protostellar mass and the outer radius of the Keplerian disk are assumed to be 0.1 $M_\odot$ and 500 au, respectively. The constant scale height of 50 au is assumed. The FWHM value of the Gaussian beam is 0.″5 (50 au).

In the integrated intensity (moment 0) maps, the distributions seem to be compact and concentrated around the protostar. Since the density of the gas is assumed to be proportional to $r^{-1.5}$, the contributions from the vicinity of the protostar are dominant. In the maps of the velocity field (moment 1 maps), the rotation motion is clearly shown. No skewed feature is seen in the moment 1 maps. In the PV diagrams along the envelope (P.A. 90°), the spin-up feature can be confirmed, except for the model with a face-on configuration. No counter velocity component which is seen in the infalling-rotating envelope model is seen in the Keplerian model, because there is no infall motion. In the PV diagrams along the direction perpendicular to the envelope (P.A. 180°), no velocity gradient is seen regardless of i. High velocity components seen in the panels for i of 30° and 60° in Fig. 3.11d are the contamination of the rotation motion near the protostar due to the finite beam size.

3.4 Outflow Model

As well as the disk/envelope system, the kinematic structure of outflows are investigated. The parabolic outflow model from [4] is employed to analyze the observed geometrical and kinematical structures of the outflow. This is just a morphological model. This parabolic model is widely applied to various low-mass and high-mass protostellar sources (e.g. [1, 2, 5, 14, 15, 17, 18]).

In this model, the outflow cavity is assumed to have a parabolic shape and its velocity is proportional to the distance to the protostar. The shape of the outflow cavity wall and the velocity of the gas on the cavity wall are represented as follows:

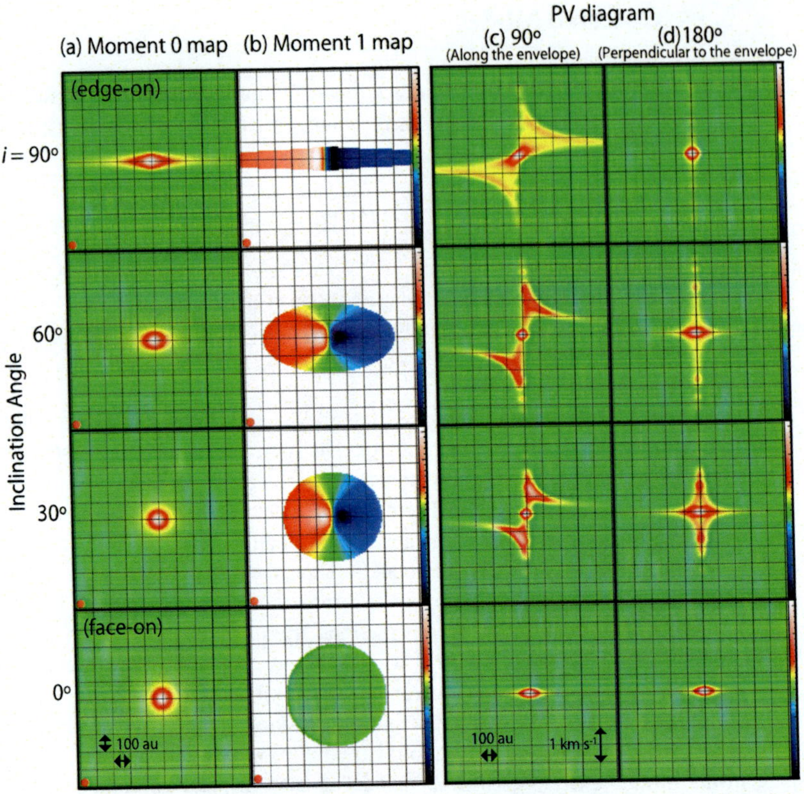

Fig. 3.11 **a** Integrated intensity maps, **b** maps of the velocity field, and **c**, **d** PV diagrams of the Keplerian model. The position axes in panels **(c)** and **(d)** are as the same as those in Fig. 3.10. The physical parameters are as follows; $M = 0.1\ M_\odot$, $R = 500$ au, and $H = 50$ au. The intrinsic line width and the beam size are 0.2 km s^{-1} and 0.″5, respectively, which are the same as those in Figs. 3.5, 3.6, 3.7, 3.8 and 3.9

$$z = CR^2, \quad v_R = v_0 \frac{R}{R_0}, \quad v_z = v_0 \frac{z}{z_0}, \tag{3.7}$$

where the z axis is taken along the outflow axis with an origin at the protostar, and R denotes the radial size of the outflow cavity perpendicular to z-axis. R_0 and z_0 are normalization constants, and both are set to be 1 au. C and v_0 are free parameters. Thus, the outflow cavity wall has a parabolic shape in this model, and it is linearly accelerated as the distance from the protostar along the outflow axis (z) and that from the outflow axis (R). In this model, the distribution of the gas is artificially assumed to be uniform for all data points, where the mass of the outflowing gas is not conserved along z-axis. An optically thin condition is also assumed, where the intensity of the line emission is proportional to the column density along the line of sight. These

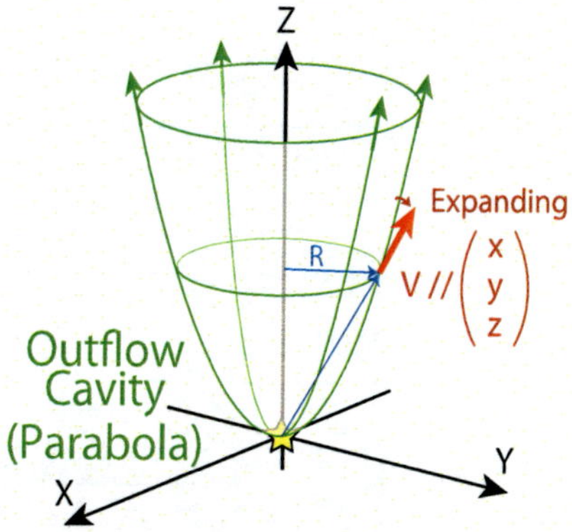

Fig. 3.12 Schematic illustration of the parabolic outflow model

simplified assumptions are employed in this study, because the main interest in the model analysis is on the velocity profile. Since the extended component would be resolved-out, it is difficult to derive accurate density profile from the observations with the interferometer. Therefore, the emissivity is not considered in the outflow analysis.

Figure 3.12 shows the schematic illustration of this model. Figure 3.13 shows how sensitive to the physical parameters the model results are in their integrated intensity map (moment 0 map), the map of the velocity field (moment 1 map), and the PV diagrams. In these models, the outflow axes are assumed to be along the north-south axis. The two outflow lobes blow in parallel to the plane of the sky with i of 90°, while they are nearly pole-on for the case at i of 30°. In each PV diagram, two parabolic features are seen. One parabolic feature corresponds to one outflow lobe. With smaller C, the parabolic features show larger opening angles. For the case at i of 90°, the gas on the outflow cavity wall shows both a red- and blue-shifted velocity at one position. These are the contributions from two positions on the cavity wall in front of and behind the outflow axis. These velocity shifts have the same absolute value. With i of 30° and 60°, the symmetric axis of the parabolic feature is red- and blue-shifted in the northern and southern sides of the protostar, respectively. Thus they can be called as the 'red-shifted lobe' or 'blue-shifted lobe'. The sign of the velocity shift is determined by the combination of the inclination angle and the curvature of the lobes. As a result, there are some parts showing a counter velocity-shift to the symmetric axis. Some parts of a lobe can appear in the counter side to the outflow axis with respect to the protostellar position. It can be confirmed as a parabolic feature which crosses the protostellar position in the PV diagram.

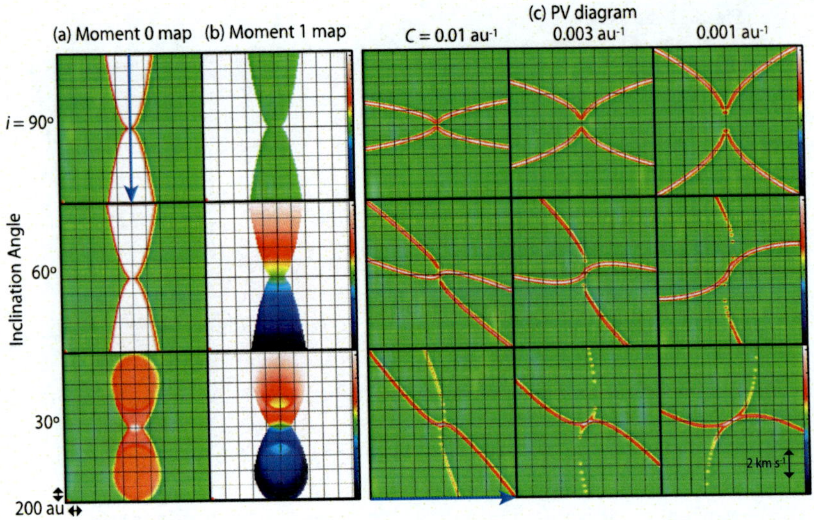

Fig. 3.13 **a** Integrated intensity maps and **b** maps of the velocity field of the outflow model with various i. The parameters of the curvature (C) and the velocity (v_0) are fixed to be 0.01 au^{-1} and 0.5 km s^{-1}. **c** PV diagrams of the outflow model along the outflow axis with various i and the curvature C (au^{-1}). The position axis is along the blue arrow shown in panel (**a**) for i of 90°. The parameter of the velocity (v_0) is fixed to be 0.5 km s^{-1}, where the velocity is 0.5 km s^{-1} at the distance of 1 au from the protostar

3.5 Physical Parameters of the Models

In the models described in Sects. 3.2–3.4, there are some key free parameters while some physical parameters are fixed. They are summarized in Table 3.1.

3.6 Examples of the Model Analysis

3.6.1 *The L1527 Case*

The infalling-rotating envelope model explained in Sect. 3.2 was first applied for L1527 [10]. L1527 is a low-mass protostellar core in Taurus (d = 137 pc [16]). The details of L1527 are described in Chap. 4.

[10] reported the high angular-resolution observations toward L1527 with ALMA. Figure 3.14a shows the integrated intensity maps of CCH and c-C_3H_2 toward L1527. The envelope component is seen to be extended along the north-south axis. Figure 3.14b shows the position-velocity (PV) diagram of CCH along the north-south axis centered at the protostellar position represented by the white cross

Table 3.1 Free and fixed physical parameters in the models

Physical parameter	Free or Fixed
Infalling-Rotating Envelope Model	
Distance (d)	Fixed (See Table 1.1)
Protostellar mass (M)	Free
Inclination angle (i)	Fixed (The value from references or derived from the outflow analysis)
Radius of the centrifugal barrier (r_{CB})	Free
Outer radius (R)	Free
Scale height of the envelope	Fixed (Uniformed or proportional to the distance from the protostar)
Beam size	Fixed based on the observations
Intrinsic line width	Fixed
Emissivity[a]	Fixed ($\propto r^{-1.5}$)
Mesh size	Fixed
Keplerian Model	
Distance (d)	Fixed (See Table 1.1)
Protostellar mass (M)	Fixed (The value derived from the envelope analysis)
Inclination angle (i)	Fixed (The same value in the envelope analysis)
Outer radius of the disk (R)	Fixed to be r_{CB}
Beam size	Fixed based on the observations
Intrinsic line width	Fixed
Emissivity[a]	Fixed ($\propto r^{-1.5}$)
Mesh size	Fixed
Outflow Model	
Distance (d)	Fixed (See Table 1.1)
Inclination angle (i)	Free (0° for a pole-on configuration)
Curvature (C)	Free
Velocity (v_0)	Free
Beam size	Fixed based on the observations
Intrinsic line width	Fixed
Emissivity[b]	Fixed to be uniform
Mesh size	Fixed

[a]In reality, the intensity of the line emission from each data point is determined by the combination of several physical conditions; e.g. the H_2 density ($n(H_2)$), the abundance ratio of the molecule to H_2 (f(X)), the gas temperature. In this model, the intensity is simply assumed to be proportional to the column density, where $n(H_2)$ is assumed to be proportional to $r^{-1.5}$ (an infalling envelope; e.g. [3, 6, 13]), f(X) is constant, and the temperature does not affect the emissivity

[b] In this chapter, the density of the molecule in the outflow model is simply assumed to be constant for all data points, because only the velocity structure is focused on in this study. In the actual analyses of the observational data (Chaps. 4, 5 and 8), only the outline of the outflow model is considered

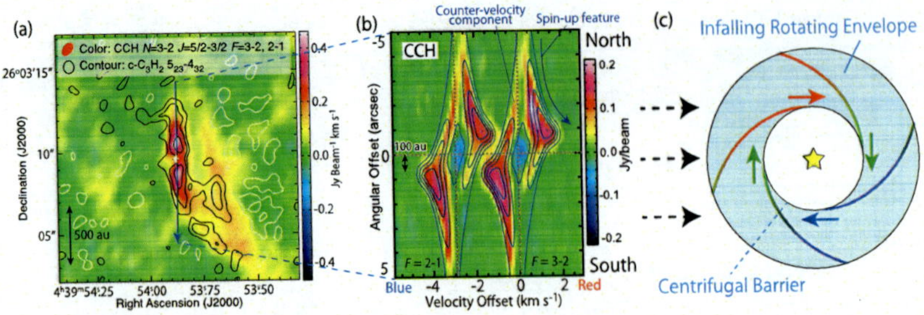

Fig. 3.14 **a** Integrated intensity maps of the CCH ($N = 3-2$, $J = 5/2-3/2$, $F = 3-2$ and $2-1$; color) and c-C_3H_2 ($5_{23}-4_{32}$; contours) lines. Contours levels are −6, −3, 3, 6, and 12 σ, where σ is 6.5 mJy beam^{-1} km s^{-1}. White contours represent a negative intensity. The white cross represents the position of the 0.8 mm continuum peak. **b** Position-velocity diagram of the CCH line prepared along the blue arrow shown in panel (**a**), along which the mid-plane of the disk/envelope system is extended. The hyperfine component ($F = 2-1$) is seen with the velocity offset of −2.8 km s^{-1}, and the same structure is repeated twice. Blue contours represent the results of the infalling-rotating envelope model. The physical parameters for the model are as follows; the protostellar mass $M = 0.18\ M_\odot$, the radius of the centrifugal barrier r_{CB} = 100 au, and the inclination angle $i = 85°$ (0° for a face-on configuration). Contour levels are every 20% of the peak intensity. **c** Schematic illustration of the infalling-rotating envelope model. Panels (**a**) and (**b**) are taken from [10] with a slight modification. © AAS. Reproduced with permission

in Fig. 3.14a. The velocity of the gas is red- and blue-shifted at the northern and southern sides of the protostar, respectively. A clear spin-up feature is seen in the PV diagram; the velocity shift increases as approaching to the protostar. The CCH emission abruptly disappears at the distance of 100 au from the protostar. There are also seen the components with the inverse velocity of the spin-up feature ('*counter-velocity component*'). The results of the infalling-rotating envelope model are represented as blue contours in Fig. 3.14b. The physical parameters for the infalling-rotating envelope model are as follows; the protostellar mass is 0.18 $M_\odot$, the radius of the centrifugal barrier is 100 au, and the inclination angle of the disk/envelope system is 85° (0° for a face-on configuration). The model results well reproduce the observed kinematic structures of the CCH lines, including the counter-velocity component.

3.6.2 The TMC−1A Case

Another example is the TMC−1A case. This source is located in the Heiles Cloud 2 ($d = 137$ pc [16]), which harbors the low-mass Class I protostar IRAS 04365+2535. It shows the chemical characteristic of WCCC similar to L1527 [12]. [9] conducted high angular-resolution observations of the CS ($J = 5-4$) and SO ($J_N = 7_6-6_5$) lines toward this source with ALMA (Fig. 3.15). The kinematic structure in the vicinity of the protostar can be explained by the above infalling-rotating envelope

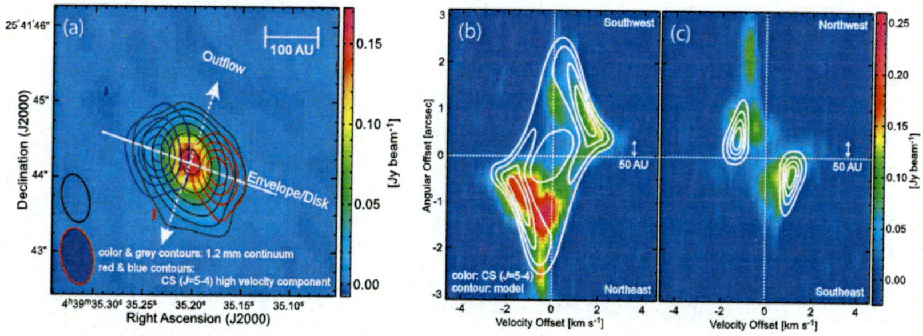

Fig. 3.15 **a** 1.2 mm continuum map (color, gray contours) and the integrated intensity maps of the high velocity components of CS ($J = 5 - 4$; red and blue contours) in TMC−1A. **b, c** PV diagrams of CS ($J = 5 - 4$) along the envelope direction (**b**) and the line perpendicular to it (**c**). The position axes are represented by the white line and arrow in panel (**a**). Taken from [9]. © AAS. Reproduced with permission

model (Fig. 3.15b), where the physical parameters are M of 0.25 $M_\odot$ and $r_{\rm CB}$ of 50 au. Figures 3.15 show the PV diagrams of CS prepared along the envelope direction and the line perpendicular to it. Although some parts of the molecular distributions seem to be missing due to the asymmetric gas distribution, the results of the infalling-rotating envelope model shown in white contours reasonably explain the kinematic structures. Thus, the infalling-rotating envelope model is expected to be a powerful tool to examine the essential kinematics in the disk forming region.

3.6.3 Some Caveats for the Model

As demonstrated in the L1527 case (Fig. 3.14), the observed kinematic structures are investigated by comparing with the physical models. The infalling-rotating envelope model is a quite simplified one as described in Sect. 3.2.1: the model does not consider any excitation effects, radiative transfer effects, and abundance variations of molecules. However, in reality, some parts of the envelope gas can be optically thick, and the distribution itself can be asymmetric around the protostar. Thus, it is not fruitful to make a fine tuning of the model so as to better match with the observed intensity. Therefore, the fundamental characteristics of the kinematic structure is focused on in this thesis. Hence, model simulations are conducted with a wide range of physical parameters, and their reasonable ranges are basically evaluated by eye. In model analyses, chi-squared tests are often employed to derive reliable values for the model parameters [8]. However, chi-squared tests for the infalling-rotating envelope model are not conducted in this thesis, considering the overwhelming systematic errors caused by the simplified assumptions described above.

References

1. Arce HG, Mardones D, Corder SA, Garay G, Noriega-Crespo A, Raga AC (2013) ALMA observations of the HH 46/47 molecular outflow. ApJ 774(1):39
2. Beuther H, Schilke P, Gueth F (2004) Massive molecular outflows at high spatial resolution. ApJ 608(1):330–340
3. Harvey DW, Wilner DJ, Myers PC, Tafalla M, Mardones D (2003) Inner structure of protostellar collapse candidate B335 derived from millimeter-wave interferometry. ApJ 583(2):809–818
4. Lee C-F, Mundy LG, Reipurth B, Ostriker EC, Stone JM (2000) CO outflows from young stars: confronting the jet and wind models. ApJ 542(2):925–945
5. Lumbreras AM, Zapata LA (2014) SMA submillimeter observations of HL Tau: revealing a compact molecular outflow. AJ 147(4):72
6. Ohashi N, Hayashi M, Ho PT, Momose M (1997) Interferometric imaging of IRAS 04368+2557 in the L1527 molecular cloud core: a dynamically infalling envelope with rotation. ApJ 475(1):211–223
7. Oya Y, Sakai N, Sakai T, Watanabe Y, Hirota T, Lindberg JE, Bisschop SE, Jørgensen JK, van Dishoeck EF, Yamamoto S (2014) A substellar-mass protostar and its outflow of IRAS 15398–3359 revealed by subarcsecond-resolution observations of H_2CO and CCH. ApJ 795(2):152
8. Oya Y, Yamamoto S (2020) Substructures in the disk-forming region of the class 0 low-mass protostellar source IRAS 16293–2422 source A on a 10 au scale. ApJ 904(2):185
9. Sakai N, Oya Y, López-Sepulcre A, Watanabe Y, Sakai T, Hirota T, Aikawa Y, Ceccarelli C, Lefloch B, Caux E, Vastel C, Kahane C, Yamamoto S (2016) Subarcsecond analysis of the infalling-rotating envelope around the class I protostar IRAS 04365+2535. ApJ 820(2):L34
10. Sakai N, Oya Y, Sakai T, Watanabe Y, Hirota T, Ceccarelli C, Kahane C, Lopez-Sepulcre A, Lefloch B, Vastel C, Bottinelli S, Caux E, Coutens A, Aikawa Y, Takakuwa S, Ohashi N, Yen H-W, Yamamoto S (2014) A chemical view of protostellar-disk formation in L1527. ApJ 791(2):L38
11. Sakai N, Sakai T, Hirota T, Watanabe Y, Ceccarelli C, Kahane C, Bottinelli S, Caux E, Demyk K, Vastel C, Coutens A, Taquet V, Ohashi N, Takakuwa S, Yen H-W, Aikawa Y, Yamamoto S (2014) Change in the chemical composition of infalling gas forming a disk around a protostar. Nature 507(7490):78–80
12. Sakai N, Sakai T, Hirota T, Yamamoto S (2008) Abundant carbon-chain molecules toward the low-mass protostar IRAS 04368+2557 in L1527. ApJ 672(1):371–381
13. Shu FH (1977) Self-similar collapse of isothermal spheres and star formation. ApJ 214:488–497
14. Takahashi S, Ho PT (2012) The discovery of the youngest molecular outflow associated with an intermediate-mass protostellar core, MMS-6/OMC-3. ApJ 745(1):L10
15. Takahashi S, Ohashi N, Bourke TL (2013) Direct imaging of a compact molecular outflow from a very low luminosity object: L1521F-IRS. ApJ 774(1):20
16. Torres RM, Loinard L, Mioduszewski AJ, Rodríguez LF (2007) VLBA Determination of the distance to nearby star-forming regions. II. Hubble 4 and HDE 283572 in Taurus. ApJ 671(2):1813–1819
17. Yeh SC, Hirano N, Bourke TL, Ho PT, Lee CF, Ohashi N, Takakuwa S (2008) The CO molecular outflows of IRAS 16293-2422 probed by the submillimeter array. ApJ 675(1):454–463
18. Zapata LA, Arce HG, Brassfield E, Palau A, Patel N, Pineda JE (2014) A spider-like outflow in Barnard 5 - IRS 1: the transition from a collimated jet to a wide-angle outflow? MNRAS 441(4):3696–3702

Chapter 4
L1527

4.1 Introduction

As demonstrated in Chap. 3, it has been found that the kinematic structure of the envelope in a low-mass Class 0/I protostellar core L1527 is well reproduced by a simple model assuming the ballistic motion. Its kinematic structure is further investigated in this chapter. The structure of the outflow cavity wall is investigated as well as the envelope component, because they are expected to be deeply related with each other.

IRAS 04368+2557 in L1527 is a representative Class 0/I low-mass protostar in Taurus (d = 137 pc [35]). Its bolometric luminosity is 1.7 $L_\odot$ [6]. It is also known as a prototypical warm carbon-chain chemistry (WCCC) source [25–27]. This source has a flattened infalling envelope with an edge-on configuration extending from north to south. [19, 42] reported the infall motion of the envelope gas conserving angular momentum based on interferometer observations. Meanwhile, [20, 31] suggested the existence of a Keplerian disk based on the observations of the ^{13}CO and $C^{18}O$ ($J = 2 - 1$) lines, respectively. Recently, a clear infalling-rotating motion in its envelope at a resolution of 0$''\!\!.$6 with ALMA was presented by [23, 24]. The observed kinematic structure was well reproduced by a simple ballistic model, and its centrifugal barrier was identified at a radius of 100 au. Moreover, a drastic change in chemical compositions across the centrifugal barrier was discovered [23]. Carbon-chain molecules and CS mainly reside in the infalling envelope outside the centrifugal barrier. Meanwhile, SO and probably CH_3OH are enhanced at the centrifugal barrier, and they may survive inside it at least partly. Such a chemical change at the centrifugal barrier had not been anticipated before.

Molecular outflows from the protostar in L1527 are extended toward the east-west direction, which is almost perpendicular to the mid-plane of the flattened edge-on envelope gas. The outflows blow almost in parallel to the plane of the sky. The outflow

---

This chapter has been published in Oya et al., 2015, ApJ, 812, 59. © AAS. Reproduced with permission.

Y. Oya, *A Few Tens au Scale Physical and Chemical Structures Around Young Low-Mass Protostars*, Springer Theses, https://doi.org/10.1007/978-981-19-1708-0_4

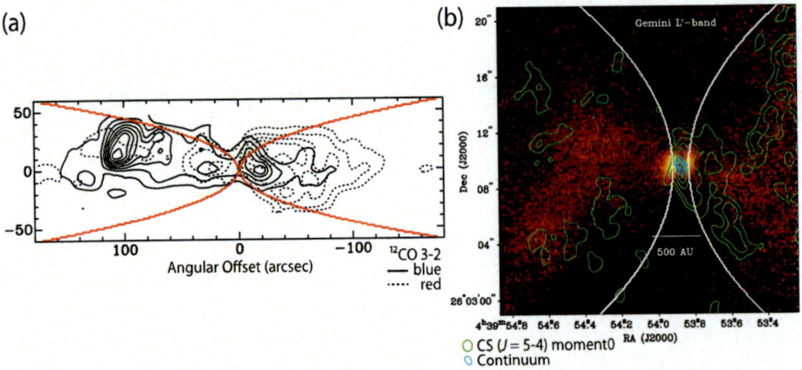

Fig. 4.1 **a** ^{12}CO map (black contours) taken from [9]. The FWHM (full width at half maximum) beam size is 11″ and the contours are drawn at the 3σ level. **b** L' band observation (color) by [33], the integrated intensity map of CS ($J = 5 - 4$; green contours) and the 1.3 mm continuum map (light blue contours). Contours for CS are 3, 6, 12, 18, and 24σ, where the rms level is 9 mJy beam^{-1} km s^{-1}. Contours for the continuum are every 20% of the peak intensity, which is 0.308 Jy beam^{-1}. Red lines in the panel (**a**) and white lines in the panel (**b**) represent the best-fit model of the outflow (Sect. 4.2.3). The parameters for the outflow model are; the inclination angle (i) is $+85°$, the curvature ($C_{\rm as}$) is 0.05 arcsec^{-1} and the velocity ($v_{\rm as}$) is 0.10 km s^{-1}. The origins of the bipolar lobes have an offset of 0.″62 from the protostellar position [33]

extending over 2′ ($\sim 2 \times 10^4$ au) scale was detected in the ^{12}CO ($J = 3 - 2$) line observed with the James Clerk Maxwell Telescope (JCMT) [9] (Fig. 4.1a). The blue-shifted and red-shifted components are prominent in the eastern and western sides of the protostellar position, respectively. As for a smaller scale, [30, 33] conducted L' band imaging of L1527 with the Gemini North telescope and the 3.6 μm band with the Infrared Array Camera (IRAC) on the *Spitzer* Space Telescope (Fig. 4.1b). Their observations delineated the butterfly shape of the outflow cavity at a ($10^3 - 10^4$) au scale. The eastern side of the outflow cavity is brighter than its western side, and thus, they suggested that the eastern cavity would point to us. If this is the case, the eastern lobe corresponds to the blue-shifted one. This result is consistent with the orientation of the larger scale outflow observed with the ^{12}CO line emission mentioned above. The protostellar-core model has been discussed by assuming this orientation of the outflow so far.

In the analysis of the ALMA observational data of the envelope gas in L1527, it has fortunately found that the orientation of its outflow/envelope system is opposite to that described above. The inclination angle of the outflow is determined and its morphological properties are characterized in this chapter.

4.2 Results

The observation was curried out as a part of the ALMA Cycle 0 operation (#2011.0.00604S). The details for this observation was reported by [24]. The observed lines are summarized in Table 4.1. The primary beam (HPBW) in this observation is 24.″5, and the synthesized beam size is 0.″8 × 0.″7 (P.A. = −6°) for the target molecular lines.

4.2.1 *Overall Distribution*

Figure 4.2 shows the integrated intensity map of the CS ($J = 5 - 4$) line. As reported in other molecular lines [23, 24], the CS line shows the envelope component extending almost along the north-south axis. Although the CS emission is heavily resolved out in the outer part in Fig. 4.2, a part of the butterfly-shaped outflow cavity wall [33] can be seen in addition to the envelope component; it is mainly on the western side of the protostar. The high density gas traced by the CS line is distributed asymmetrically around the protostar.

The position-velocity (PV) diagram along the outflow axis is shown in Fig. 4.3. Envelope components are concentrated at the protostellar position, while outflow components are extended along the east-west axis from there. The outflow seems to be accelerated as the distance from the protostar. The velocity shifts of the outflow are almost symmetric to the protostellar position, although blue-shifted components are prominent on the western side of the protostar. Therefore, it is obvious that the outflow blows almost in parallel to the plane of the sky, which is consistent with previous reports [30, 33]. Because the outflow components are faint and heavily resolved out, it is difficult to derive the inclination angle accurately from the kinematic structure of the outflow. Hence, the envelope components are first investigated to characterize the geometry of this source.

Table 4.1 Parameters of the Observed Line[a]

Molecule	Transition	Frequency (GHz)	$E_u k_B^{-1}$ (K)	$S\mu^{2\,b}$ (D^2)
CS	$J = 5 - 4$	244.9355565	35	19.17
c-C_3H_2	$5_{2,3} - 4_{3,2}$	249.0543680	41	76.32

[a]Taken from CDMS [18]

[b]Nuclear spin degeneracy is not included

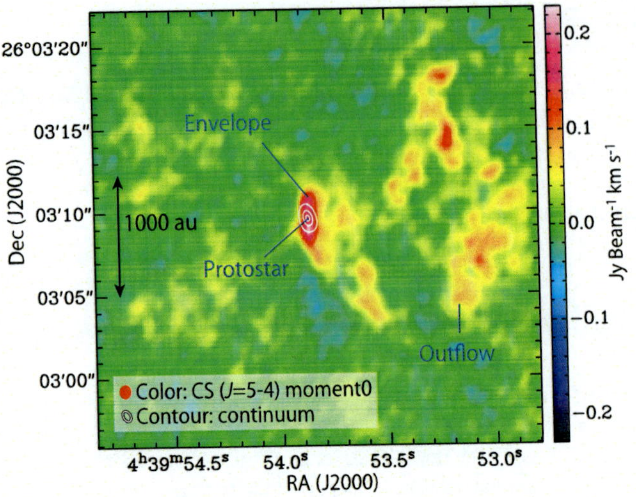

Fig. 4.2 Integrated intensity map of the CS ($J = 5-4$; color) line and the 1.3 mm continuum map (white contours). The velocity range for the integration is from 3.1 to 8.7 km s^{-1}. Contour levels for the continuum are 30, 150, and 270σ, where the rms level is 1 mJy $beam^{-1}$

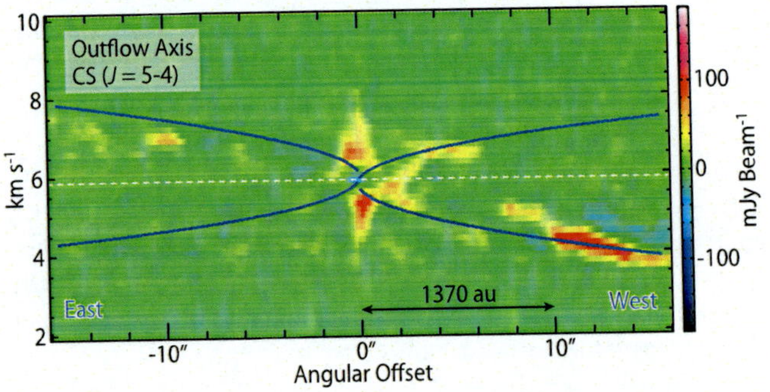

Fig. 4.3 Position-velocity (PV) diagram of the CS ($J = 5-4$; color) line prepared along the outflow axis. The blue lines represent the result of the outflow model (Sect. 4.2.3) with the parameters of $i = +85°$, $C_{as} = 0.05$ $arcsec^{-1}$, and $v_{as} = 0.10$ km s^{-1}. In this model, no offset of the origin of the outflow from the protostar is assumed

4.2.2 Envelope

The kinematic structure of the protostellar envelope of L1527 observed in the c-C_3H_2, CCH, and CS emission is well reproduced with a model of an infalling-rotating envelope [23, 24]. Details of this model are described in Chap. 3. In this model, the particles cannot fall inward of a certain radius (perihelion) due to the conservation of angular momentum and energy. This radius defines the centrifugal barrier. The velocity field of the particle motion in this ballistic model is characterized

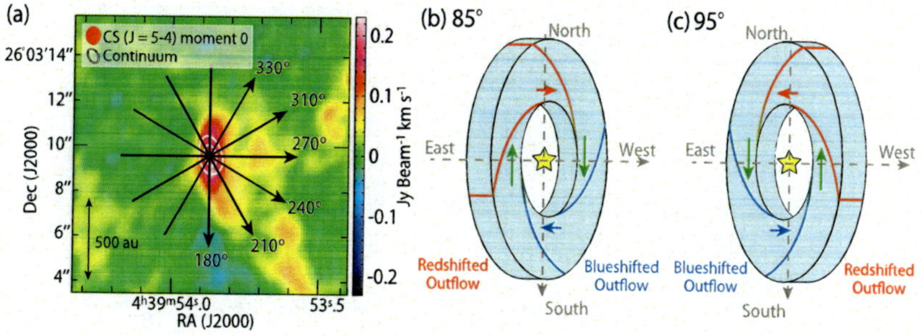

Fig. 4.4 **a** A blow-up of the central part of Fig. 4.2; the integrated intensity map of the CS ($J = 5-4$; color) line and the continuum map (white contours). Contours for continuum are as the same as those in Fig. 4.2. Black arrows represent the lines along which the PV diagrams in Figs. 4.5, 4.6 and 4.7 are prepared. **b**, **c** Schematic illustrations of the envelope model with the inclination angle of (**b**) 85° and (**c**) 95°

by the protostellar mass and the radius of the centrifugal barrier. The PV diagram of the model prepared along the mid-plane of the envelope does not depend on the direction of its inclination angle from the plane of the sky but on its absolute value. Hence, [23] (Chap. 3) did not consider the direction of the inclination with respect to the edge-on configuration ($i = 90°$) in the analysis of the PV diagram. Therefore, the PV diagrams of the CS ($J = 5-4$) and c-C_3H_2 ($5_{2,3} - 4_{3,2}$) lines are examined, which are prepared along various lines passing through the protostellar position, to investigate the direction of the inclination angle of the envelope gas.

The PV diagrams of the CS ($J = 5-4$) line along the six lines with different position angles (Fig. 4.4a) are shown in Figs. 4.5 and 4.6. In these figures, the ballistic models have the inclination angle (i) of 85° and 95°, which correspond to the cases shown in Figs. 4.4b, c, respectively. It should be noted that an inclination angle of 95° is assumed in [32], where the eastern side of the flattened envelope faces to us (Fig. 4.4c). The other physical parameters employed for the models are described in Appendix of this chapter. The CS emission near the centrifugal barrier is more enhanced than in the outer part of the envelope component in comparison with the c-C_3H_2 ($5_{2,3} - 4_{3,2}$) case [24]. Moreover, the red-shifted components are slightly weaker than the blue-shifted ones, and the self-absorption effect is seen for the systemic velocity (5.9 km s^{-1} [26]). Nevertheless, the kinematic structure along the mid-plane of the envelope (P.A. "180°") is well explained by the above model assuming the either direction of the inclination (i = 85° or 95°). Since the compact distribution around the protostar is focused on for the envelope analysis, resolved-out components are negligible here.

It is evident that the model with the inclination angle of 85° well reproduces the PV diagrams along all the lines (Fig. 4.5). On the other hand, the model with the inclination angle of 95° does not (Fig. 4.6). In particular, the PV diagrams labeled as "240°", "270°", and "300°" in Figs. 4.5 and 4.6, which represent the infall motion rather than the rotation motion, seem to be reproduced better by the model with the inclination angle of 85° (Fig. 4.5) than that with 95° (Fig. 4.6). For instance, the

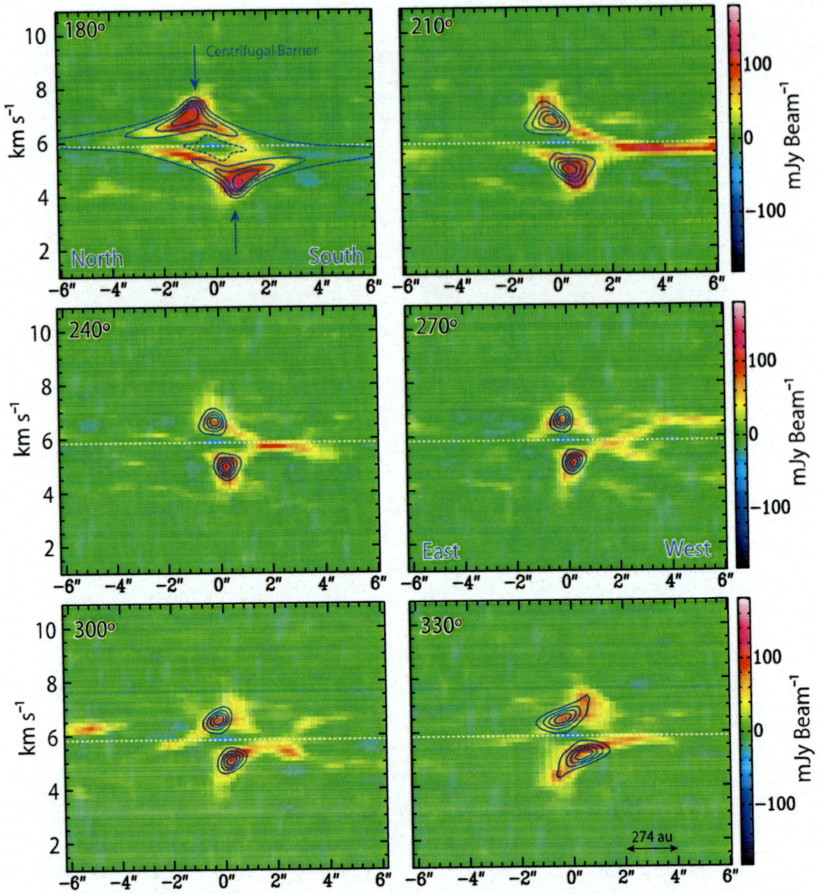

Fig. 4.5 PV diagrams of the CS ($J = 5 - 4$; color) line along the lines shown in Fig. 4.4a. The label in the left upper corner ("180°", "210°", "240°", "270°", and "330°", and "330°") represents the position angle of the position axis in each panel. Blue contours represent the infalling-rotating envelope model with the inclination angle of 85° (Fig. 4.4b). Contour levels are every 20 % of the peak intensity in each panel. The dashed contour around the central position in the panel "180°" represents the dip toward the center

observed PV diagram along the "270°" line shows two peaks; one at the eastern side and the other at the western side of the protostar with the red-shifted velocity and the blue-shifted velocity, respectively. The model with the inclination angle of 85° reproduces this trend, while that with 95° does not; in the latter case, the intensity peak at the eastern side shows the blue-shifted velocity, while the other peak at the western side shows the red-shifted velocity. Thus, the observational results likely prefer the model with the inclination angle of 85°.

In order to evaluate the goodness of the fit more quantitatively, optimization of the inclination angle is attempted based on the fitting for the observed PV diagrams. For

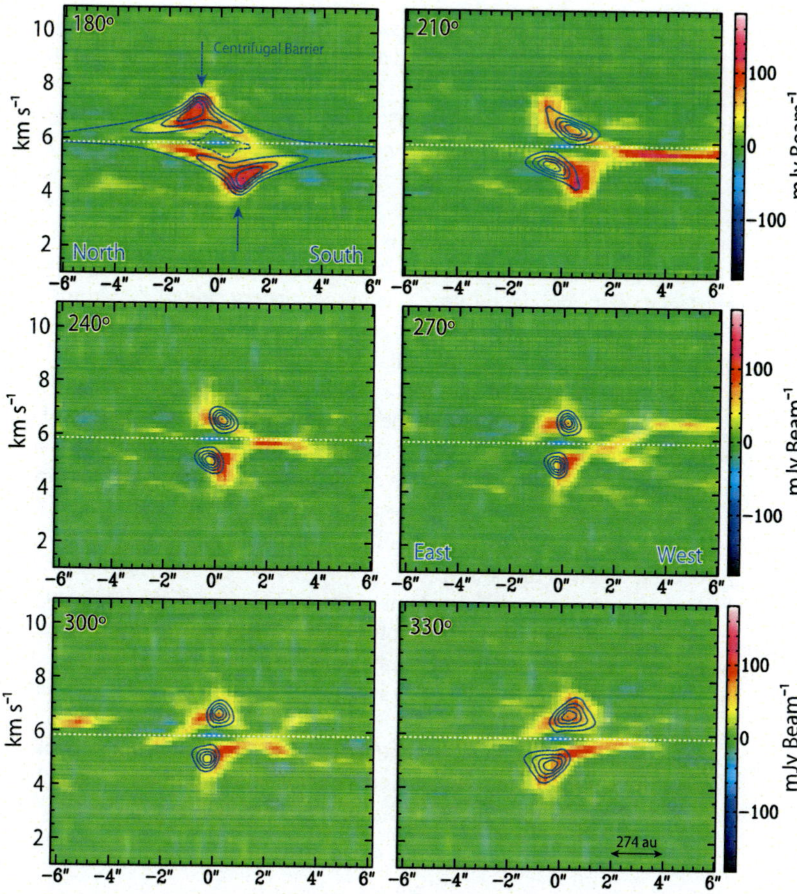

Fig. 4.6 PV diagrams of the CS ($J = 5 - 4$; color) line along the lines shown in Fig. 4.4a. The position axes are taken as those in Fig. 4.5. Blue contours represent the infalling-rotating envelope model with the inclination angle of 95° (Fig. 4.4c). Contour levels are every 20 % of the peak intensity in each panel. The dashed contour around the central position in the panel "180°" represents the dip toward the center

this purpose, model calculations are conducted with the other inclination angles; the results are shown in Appendix of this chapter. The model used here is a simplified one involving many assumptions described in Appendix of this chapter in order to explore just the basic physical and kinematic structure of the envelope. Thus, the fitting suffers from systematic errors due to these assumptions, and the statistical argument based on the root-mean-square (rms) of the difference between PV diagrams is almost meaningless. Nevertheless, it is found that the observational results are reasonably reproduced by the model with the inclination angle from 80° to 85°. More conservatively, it is constrained to be larger than 75°. Moreover, it is firmly concluded that the inclination angle is less than 90°; the positions and the velocities of the two intensity peaks mentioned above are not consistent with the observations, if the opposite direction, where the eastern surface of the envelope faces to us, are employed for the model. [30] reported that the morphology of the L' band image cannot be explained, if the disk is inclined by much smaller than 80°. Hence, the inclination angle obtained above is consistent with the previous reports except for the direction of the inclination angle. With this in mind, the inclination angle of 85° is employed in the following analysis.

With the inclination angle less than 90°, the western side of the envelope component faces to us, and the outflow axis in the western side of the protostar points to us, as shown in Fig. 4.4b. This configuration is opposite to that reported previously [9, 33]. On the other hand, it is consistent with the velocity map (moment 1 map) of the SO ($J_N = 7_8 - 6_7$) line reported by [24]. The velocity map of the SO line shows a slightly skewed feature; the red-shifted component slightly extends from the northern part to the eastern part due to the rotation and infall motion, while the blue-shifted one slightly extends from the southern part to the western part. This indicates that the eastern part and the western part of the envelope are in front of and behind the protostar, respectively (Fig. 4.4b). This configuration is consistent with the interpretation in this study.

Figure 4.7 shows the PV diagrams of the c-C_3H_2 ($5_{2,3} - 4_{3,2}$) line prepared along the six lines passing through the protostellar position shown in Fig. 4.4a. As well as the CS emission, the red-shifted components of the c-C_3H_2 emission are weaker than the blue-shifted components, and its systemic velocity components are self-absorbed. Nevertheless, the PV diagrams of the c-C_3H_2 line seem to be well reproduced by the model with the inclination angle of 85° shown by the blue contours. Because c-C_3H_2 preferentially exists in the envelope [23], the CS emission which is not seen in the c-C_3H_2 emission would represent the outflow component.

4.2.3 *Outflow*

In Sect. 4.2.2, the inclination angle of the infalling-rotating envelope was successfully evaluated based on its kinematic structure. Then, assuming that the inclination angle of the outflow axis is perpendicular to the mid-plane of the envelope, the kinematic structure of the outflow is investigated (Fig. 4.3). A parabolic outflow model is employed to analyze the observed outflow structure (e.g. [14, 21]). The shape of the outflow cavity wall and the velocity structure in this model are represented by the following formulae [14]:

$$z = C_{\rm as} R^2, \quad v_R = v_{\rm as} \frac{R}{R_0}, \quad v_z = v_{\rm as} \frac{z}{z_0}. \tag{4.1}$$

R denotes the radial size of the cavity perpendicular to the outflow axis (z axis). R_0 and z_0 are normalization constants, both of which are set to be 1″ [21]. $C_{\rm as}$ and $v_{\rm as}$ are free parameters. Note that these parameters correspond to $(C \times D)$ and $(v_0 \times D)$ in

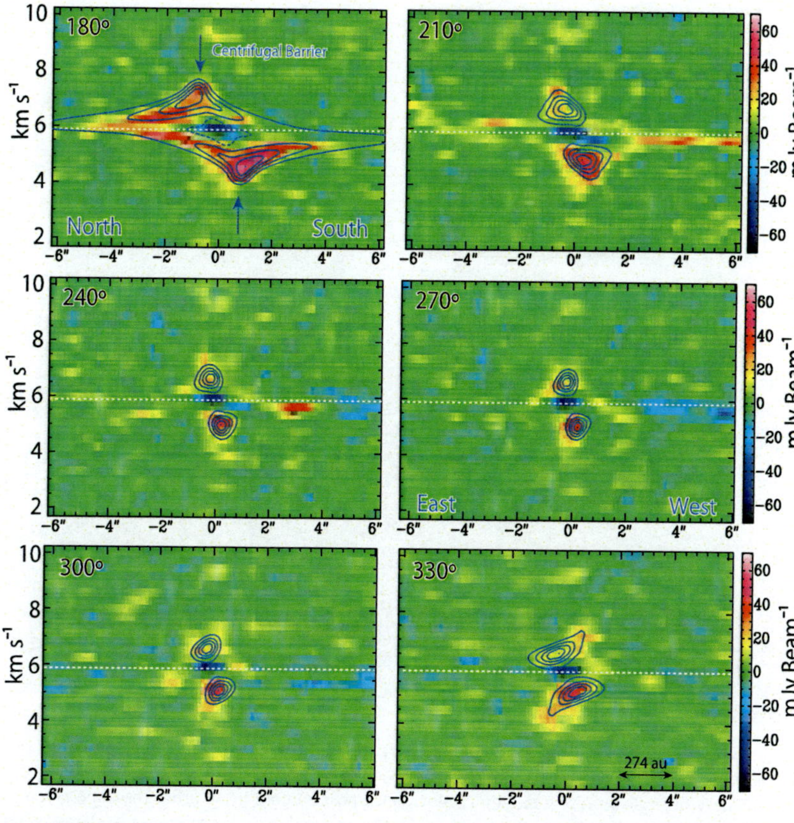

Fig. 4.7 PV diagrams of the c-C_3H_2 ($5_{2,3} - 4_{3,2}$; color) line along the lines shown in Fig. 4.4a. The infalling-rotating envelope model represented by the blue contours are as the same as that in Fig. 4.5 (i = 85°). Contour levels are every 20% of the peak intensity in the model. The dashed contour around the central position in the panel "180°" represents the dip toward the center

the outflow model described in Chap. 3, where D denotes the distance of the source from the Sun (See also Chap. 10). Thus, the wall of the outflow cavity is assumed to have a parabolic shape. The gas on the cavity wall is assumed to be linearly accelerated as the distance from the protostar. Such a parabolic model has widely been employed for various low-mass and high-mass protostellar sources (e.g. [1, 2, 16, 28, 29, 41, 44]). In this model, the intensity of the emission is assumed to be proportional to the column density, where the molecular density is simply assumed to be constant. Although these assumptions are not realistic, it does not matter the following analysis, where the velocity structure is focused on while the intensity profile is neglected.

In Fig. 4.3, the blue lines represent the best-fit model by eye. The inclination angle is fixed to 85° based on the analysis of the envelope structure (Sect. 4.2.2). The parameters for the best-fit model are; $C_{\rm as} = 0.05$ arcsec^{-1}, and $v_{\rm as} = 0.10$ km s^{-1}.

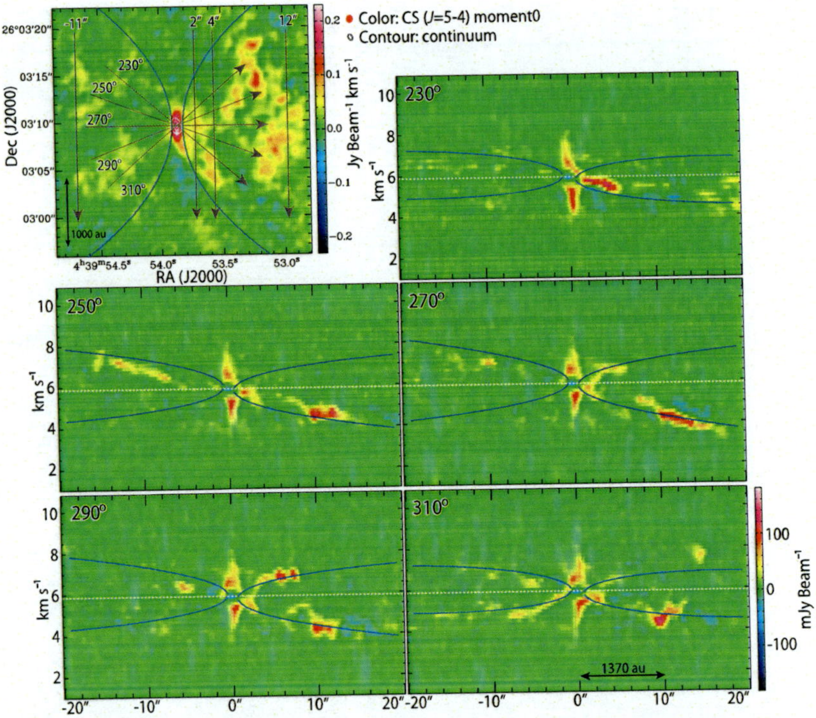

Fig. 4.8 Integrated intensity (moment 0) map (upper left panel) and PV diagrams of CS ($J = 5-4$). The integrated intensity map of CS is the same as in Fig. 4.2. White contours in the integrated intensity map represent the continuum map, whose contour levels are as the same as those in Fig. 4.2. PV diagrams are prepared along the gray arrows passing through the protostellar position shown in the integrated intensity map. The blue lines represent the result of the best-fit outflow model, where the parameters are as follows; $i = 85°$, $C_{\rm as} = 0.05$ arcsec^{-1}, $v_{\rm as} = 0.10$ km s^{-1}, and the origins of the lobes have an offset of 0$''$.62 from the protostellar position [33]

This model seems to explain the accelerated outflow component on the western side of the protostar. Meanwhile, the velocity structure in the vicinity of the protostar is not well reproduced. This result suggests that the origin of the parabolic shape has a certain offset from the protostellar position as pointed out by [30].

In the outflow model, the offset of 0$''$.62 (85 au) is adopted for the outflow origin [33]. The integrated intensity map of the CS ($J = 5-4$) line is shown in the upper left panel in Fig. 4.8. The other panels in Fig. 4.8 represent the PV diagrams of the CS line, whose position axes are prepared along the arrows depicted in the integrated intensity map. The best-fit outflow model is shown by the blue lines in the PV diagrams, where the inclination angle = 85°, $C_{\rm as} = 0.05$ arcsec^{-1}, and $v_{\rm as} = 0.10$ km s^{-1}. With the inclination angle lower than 70° or higher than 95°, the model does not well reproduce the observed kinematic structure with any values of $C_{\rm as}$ and $v_{\rm as}$. This constraint is consistent with that found in the analysis of the envelope structure (Sect. 4.2.2). The

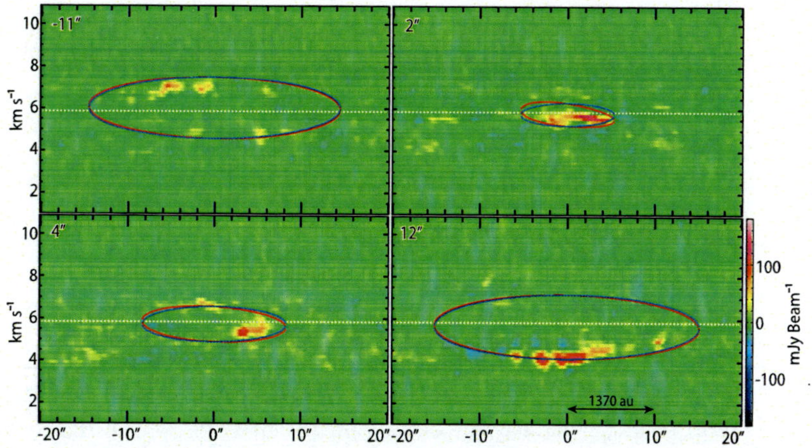

Fig. 4.9 PV diagrams of CS ($J = 5 - 4$; color) across the outflow axis, where the position axes are shown in the upper left panel in Fig. 4.8. The blue lines represent the result of the best-fit outflow model, where the parameters are; $i = 85°$, $C_{as} = 0.05$ arcsec^{-1}, and $v_{as} = 0.10$ km s^{-1}, and the origins of the lobes have an offset of 0.″62 from the protostellar position [33]. The red lines represent another outflow model, where a rotation motion of the outflow cavity wall is taken into account (See Sect. 4.3.2), although they are almost overlapped with the blue lines

best model with the inclination angle of 85° is represented in Fig. 4.1a (red lines), b (white lines), and the upper left panel in Fig. 4.8 (blue lines). Even though the parameters for the outflow model are obtained based on the kinematic structure of the outflow and envelope, this model likely explains the spatial distribution of the outflow in the vicinity of the protostar. The eastern and western outflow lobes are reported to have the maximum velocity shifts of 6.9 and 9.6 km s^{-1} from the systemic velocity (5.9 km s^{-1}[26]) based on the ^{12}CO ($J = 3 - 2$) observation [9]. In the obtained model, the two outflow lobes have the maximum velocity shift of 8.2 km s^{-1} at the distance of 200″ ($\sim$ 3000 au) from the protostar. Although the parameters for the outflow model are constrained based on the kinematic structure on a few tens of arcsecond, the model is likely consistent with the observational results on a larger scale.

Figure 4.9 represents the PV diagrams of the outflow, whose position axis is prepared along the line perpendicular to the outflow axis (the upper left panel in Fig. 4.8). The best model represented by the blue lines seems to reproduce the basic features in the observation. If the outflow blows almost in parallel to the plane of the sky, PV diagrams across the outflow axis are expected to show an elliptic feature, as seen in the IRAS 15398−3359 case (Chap. 5). Such an elliptic feature is partially seen in the PV diagrams.

It should be noted that the eastern and western lobes may have different kinematic and geometric structure with each other, which make it difficult for the simple parabolic model to perfectly explain the observed distribution (for instance, the PV diagrams in Fig. 4.8). This may originate from asymmetry in the distribution of ambient gas. A detailed discussion on this issue is outside the scope of this thesis, since the CS ($J = 5 - 4$) distributions in the outflow are faint and heavily resolved out.

4.3 Discussion

4.3.1 *Direction of the Outflow*

In Sect. 4.2.2, the inclination angle of the infalling-rotating envelope is constrained including its direction with the aid of a ballistic model, and the obtained inclination angle is consistent with the outflow structure traced in the CS emission. However, if the outflow axis is perpendicular to the mid-plane of the envelope, the direction of the inclination contradicts pervious reports for the outflow [9, 30, 33]. This contradiction is discussed below.

The eastern surface of the envelope is assumed to face the observer by [30, 33] (Fig. 4.4c). This configuration is assumed based on their observation of the infrared (L' band) reflection by the outflow cavity wall, which is brighter on the eastern side than on the western side in the vicinity of the protostar. The contradiction of the direction of the outflow between their reports and this study can originate from the two reasons described below.

The first possibility is the extinction by inhomogeneous gas distribution around the protostar. In Fig. 4.1b, the integrated intensity map of the CS ($J = 5 - 4$) emission is represented by the green contours. A ridge structure extending along the western outflow cavity is detected in the CS emission on the southwestern side of the protostar. The L' band observation shows relatively weak emission in this ridge. This indicates that the scattered light in the L' band is obscured by the dense gas of the outflow cavity wall on its side near the observers. In fact, the bright CS emission in this ridge is slightly blue-shifted (the upper right panel in Fig. 4.9), indicating that this component is in front of the outflow axis. The column density of CS was evaluated to be $(1 - 2) \times 10^{13}$ cm^{-2} at the western position with an offset of 0$''$.7, which is the FWHM (full width at half maximum) of the beam size along the east-west direction, from the protostellar position. This value was derived by using the RADEX code [36], where the H_2 density and the kinetic temperature were assumed to be $(10^6 - 10^7)$ cm^{-3} and 30 K, respectively [23, 24]. Then, the H_2 column density is roughly estimated to be $(1 - 2) \times 10^{22}$ cm^{-2} by using a typical value for the fractional abundance of CS relative to H_2 (10^{-9}; e.g. [37, 39]). This column density of H_2 corresponds to an extinction of about $(0.5 - 1)$ mag in the L' band.

Since the observed CS emission is heavily resolved out and mainly traces dense gas ($>10^6$ cm^{-3}), the above estimation of the extinction in the L' band could be the lower limit. If the L' band emission is obscured by the outflow cavity wall, its brightness depends on the distribution of matter in the outflow cavity on the two sides of the protostar. With this situation, the direction of the inclination cannot be constrained based on the asymmetry in the observed brightness. [38] reported that the

Spitzer 3.6 μm HiRes deconvolved image shows the brightness asymmetry of the outflow cavity in the vicinity of the protostar opposite to that reported by [30, 33]. The result by [38] seems consistent with the configuration found in this analysis (Fig. 4.4b). Thus, the observations of the morphology of the cavity reflection seem inconsistent with each other. Hence, the configuration of the outflow/envelope system derived from the kinematics of the infalling-rotating envelope is more reliable than that from the brightness morphology.

The second possibility for the contradiction in the suggested outflow direction is that the outflow axis is not perpendicular to the mid-plane of the envelope structure. Some asymmetry in the protostellar source, such as its binarity, may cause this situation, although the binarity of L1527 is controversial [15]. In the studies for exoplanets, it has been reported that the spin axis of the central star and the orbital axes of planets around it are not always parallel to each other (e.g. [40]). Although the scattering among planets are considered to be an important cause for it, the angular momentum of the central star could have a different axis with that of the associated disk. This is an important issue for exoplanet studies. From this point of view, exploring this possibility would be very interesting.

According to the ^{12}CO ($J = 3 - 2$) observation reported by [9], the eastern and western outflow lobes are mainly blue-shifted and red-shifted, respectively, although both the lobes show the counter velocity components. If the outflow has a different direction between on the small scale and the large scale, it may imply the precession of the outflow axis. The dynamical time scale is reported to be 1.5×10^4 and 9.0×10^3 yr for the eastern and western lobes, respectively, without the correction for the inclination angle [9], although these values are rough estimates based on the apparent maximum velocity shifts and the lengths of the lobes. Since the outflow lobes in this source blow almost in parallel to the plane of the sky, the velocity along the line of sight (v_{LOS}) detectable in observational studies almost consists of the expanding motion of the cavity wall instead of the outflowing motion. Hence, the velocity along the outflow axis can hardly be estimated by $v_{\mathrm{LOS}}/\cos i$, where i is the inclination angle of the outflow axis (0° for the pole-on geometry). Meanwhile, the distance from the protostar along the outflow axis (z axis) is well approximated by the apparent distance from the protostar on the plane of the sky with the inclination angle of 85°. The line of sight has two intersections with each outflow cavity wall, and the interactions have different line-of-sight velocities with each other (v_{LOS}). In the red-shifted lobe, the velocity along the outflow axis (v_z) and the expanding velocity (v_r) perpendicular to it in the parabolic model are calculated to be $(v_z, v_r) = (9.6\ \mathrm{km\ s^{-1}}, 4.4\ \mathrm{km\ s^{-1}})$ and $(v_z, v_r) = (10.4\ \mathrm{km\ s^{-1}}, 4.6\ \mathrm{km\ s^{-1}})$ for the intersection in front of the outflow axis and that behind the axis, respectively, at the distance projected on the plane of the sky of 100″ from the protostar. Hence, the velocities along the line of sight (v_{LOS}) are calculated to be -3.5 and 5.4 km s^{-1} for the two intersections. The outflowing and expanding velocities in the blue-shifted lobe are calculated as well as those in the red-shifted lobe, while their velocity shifts from the systemic velocity have the opposite signs to those for the red-shifted lobe. Then the age of the lobes is estimated to be $t = z/v_z \sim 6.5 \times 10^3$ yr. Assuming the inclination angle of 85° and 95° on the 10″ and 100″ scales, respectively, a precession of 10° in $\sim 6.5 \times 10^3$ yr is speculated. Here, no other precession event is assumed in between. This precession rate is comparable to that found in L1157; [7] reported a precession of 6° in $\sim 4 \times 10^3$ yr.

Alternatively, the observed emission in the outflow cavity could be inhomogeneous; the blue-shifted and red-shifted emissions could be dominant in the eastern and western lobes on the large scale, respectively, by accident. This situation could happen because the outflow lobes blow almost in parallel to the plane of the sky.

4.3.2 Angular Momentum

In the model of the infalling-rotating envelope, it is assumed that the angular momentum conservation prevents the envelope gas from falling inward of the centrifugal barrier (Chap. 3). However, [23] pointed out that H_2CO possibly resides inside the centrifugal barrier. If this is the case, the angular momentum of the envelope gas has to be extracted so that the gas falls beyond the centrifugal barrier. The outflow is a candidate mechanism for the angular momentum extraction (e.g. [17, 34]). Hence, possible rotation motion in the outflow is worth exploring. In fact, rotation motion in outflows and jets has been suggested for various protostellar sources [3–5, 8, 10–13, 22, 43] (Chap. 8). Rotation motion is mostly reported for collimated jets, while rotation motion in a well spread outflow like the L1527 outflow is not very evident (B59#11 [8]).

The velocity field in the outflow of L1527 is almost symmetric to the outflow axis in Fig. 4.9. Thus, a rotation motion is hardly seen in this outflow. If the outflow plays a role in extraction of the angular momentum from the infalling-rotating envelope, its northern and southern edges are expected to be more red-shifted and blue-shifted than the parabolic outflow model, respectively. Another outflow model, where the rotation motion on the outflow cavity wall is considered, is shown by the red lines in Fig. 4.9. In this model for a rotating outflow, the launching point of the outflow is set to be at the centrifugal barrier, and the specific angular momentum of the outflow is assumed to be conserved [10]. The specific angular momentum of the outflow in this model is roughly approximated to be the same as that of the envelope gas. The latter is calculated based on the protostellar mass and the radius of the centrifugal barrier (Sect. 4.2.2). Then, the rotation motion in the outflow is represented by the radial size (R) of the outflow cavity wall perpendicular to the outflow axis as:

$$v_{\rm rot} = \frac{\sqrt{2GMr_{\rm CB}}}{R}, \tag{4.2}$$

where $r_{\rm CB}$ denotes the radius of the centrifugal barrier in the infalling-rotating envelope, M denotes the protostellar mass, and G is the gravitational constant. The rotation motion in the outflow would be large near its launching point, (i.e. where R is small; the upper right panel in Fig. 4.9), and decrease as the distance from the protostar (larger R; the bottom right panel in Fig. 4.9). This would be a reason why the rotation motion has been reported for well collimated jets. In Fig. 4.9, the outflow model with a rotation motion almost overlaps with the model without a rotation motion. Therefore, it is essential to investigate the kinematic structure near the launching point of the outflow at a higher spatial resolution.

In this chapter, the results obtained from the analysis of the envelope structure of L1527 was utilized to investigate the kinematic structure of the outflow. Recent high capabilities in millimeter/submillimeter-wave observations make it possible to unveil the small-scale structure of the envelope; specifically, the kinematic structure in the vicinity of the centrifugal barrier. As well as the envelope structure, it should be emphasized that the outflow structure can be investigated by using observations focused on a narrow region around the protostar with the aid of the simple parabolic model. Morphology of outflows on a large scale often suffers from interactions with the ambient gas, and thus, elimination of such an effect is indispensable for characterization of outflows, as discussed by [38]. Hence, observing the outflow structure in the vicinity of the protostar will be essential to further studies, as demonstrated in this chapter. Moreover, observations with such high angular-resolutions will delineate the inner structure of envelopes as well as the launching point of the outflows. They would be related to each other in the formation of rotationally-supported disks.

Appendix: Envelope Model with Various Inclination Angles

The PV diagrams of the CS ($J = 5 - 4$) line in Fig. 4.10 are prepared along the two lines passing through the protostellar position shown in Fig. 4.4. One of the two lines is perpendicular to the outflow axis ("180°"), while the other is parallel to it ("270°"). As well, the results of the infalling-rotating envelope model with various inclination angles are represented by the blue contours. In these models, the protostellar mass and the radius of the centrifugal barrier are fixed to be 0.18 $M_\odot$ and 100 au, respectively, which are determined based on the observational results [24]. The outer radius of the envelope is fixed to be 1000 au, which well reproduces the PV diagram of CCH (Chap. 3 [23]). The emission is convolved with a Gaussian beam with FWHM of $0''.5 \times 0''.5$ and the intrinsic line width of 0.5 km s^{-1}. In this model, the intensity of the emission is simply assumed to be proportional to the column density, where the molecular density is assumed to be proportional to $r^{-1.5}$. The effects of the optical depth, the excitation, and the temperature gradient are not taken into account. The gas kinetic temperature is reported to be (30 – 60) K for the infalling-rotating envelope component and the centrifugal barrier [23, 24]. With the assumption of the optically thin condition and the constant molecular abundance, the intensity is proportional to $\frac{1}{U(T)} \exp\left(-\frac{E_\mathrm{u}}{kT}\right)$, where T denotes the gas temperature, $U(T)$ the partition function of the upper state, E_u the upper state energy, and k the Boltzmann constant. With the upper energy of the CS ($J = 5 - 4$) line of 35 K, $\frac{1}{U(T)} \exp\left(-\frac{E_\mathrm{u}}{kT}\right)$ is calculated to be 0.012 and 0.011 with T of 30 and 60 K, respectively. Thus, the temperature gradient does not seriously affect the intensity profile in the infalling-rotating envelope.

The optically thin condition for the velocity components outside the systemic velocity (5.9 km s^{-1}[26]) can be justified for the CS ($J = 5 - 4$) line; the optical depth of this line toward the centrifugal barrier is estimated to be 0.1 with the LVG (large velocity gradient) approximation for the H_2 density range from 3×10^6 to 3×10^7 cm^{-3} and the temperature range from 30 to 60 K. Since the CS line heavily suffers from the self absorption effect at the systemic velocity, the low velocity-shift components are not considered to compare with the envelope models.

In Fig. 4.10, the panels for the inclination angle of 85° and 95° are as the same as those in Figs. 4.5 and 4.6, respectively. As discussed in Sect. 4.2.2, the PV diagram along the line perpendicular to the outflow axis (i.e. along the mid-plane of the

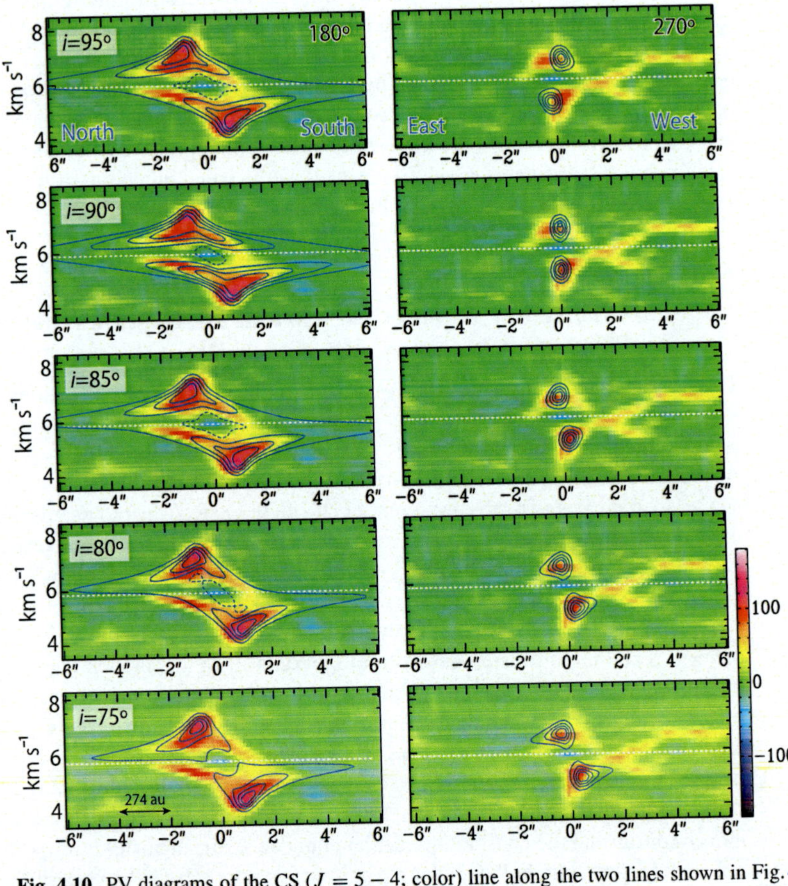

Fig. 4.10 PV diagrams of the CS ($J = 5 - 4$; color) line along the two lines shown in Fig. 4.4a. The position axes are prepared to be parallel to the mid-plane of the envelope structure ("180°") or perpendicular to it ("270°"). Blue contours in each row represent the results of the infalling-rotating envelope model with the inclination angle of 95° (Fig. 4.4c), 90° (edge-on), 85° (Fig. 4.4b), 80°, or 75°. Contour levels are every 20% of the peak intensity in each panel. The dashed contours around the central position in the panels in the left column, except for the panel for i of 75°, represent the dip toward the center. The panels for the inclination angles of 85° and 95° are the same as those in Figs. 4.5 and 4.6, respectively

envelope; "180°") depends on the absolute value of the inclination angle regardless of its direction with respect to the edge-on configuration ($i = 90°$). On the other hand, the PV diagram along the outflow axis ("270°") depends on the infall motion of the envelope gas more significantly, and thus, it is sensitive to the direction of the inclination.

References

1. Arce Héctor G, Mardones Diego, Corder Stuartt A, Garay Guido, Noriega-Crespo Alberto, Raga Alejandro C (2013) ALMA observations of the HH 46/47 molecular outflow. ApJ 774(1):39
2. Beuther H, Schilke P, Gueth F (2004) Massive molecular outflows at high spatial resolution. ApJ 608(1):330–340
3. Choi M, Kang M, Tatematsu K (2011) Rotation of the NGC 1333 IRAS 4A2 protostellar jet. ApJ 728(2):L34
4. Codella C, Cabrit S, Gueth F, Cesaroni R, Bacciotti F, Lefloch B, McCaughrean MJ (2007) A highly-collimated SiO jet in the HH212 protostellar outflow. A&A 462(3):L53–L56
5. Coffey D, Bacciotti F, Ray TP, Eislöffel J, Woitas J (2007) Further indications of jet rotation in new ultraviolet and optical hubble space telescope STIS spectra. ApJ 663(1):350–364
6. Green JD, Evans NJ, Jørgensen JK, Herczeg GJ, Kristensen LE, Lee JE, Dionatos O, Yildiz UA, Salyk C, Meeus G, Bouwman J, Visser R, Bergin EA, van Dishoeck EF, Rascati MR, Karska A, van Kempen TA, Dunham MM, Lindberg JE, Fedele D, DIGIT Team (2013) Embedded protostars in the dust, ice, and gas in time (DIGIT) herschel key program: continuum SEDs, and an inventory of characteristic far-infrared lines from PACS spectroscopy. ApJ 770(2):123
7. Gueth F, Guilloteau S, Bachiller R (1996) A precessing jet in the L1157 molecular outflow. A&A 307:891–897
8. Hara C, Shimajiri Y, Tsukagoshi T, Kurono Y, Saigo K, Nakamura F, Saito M, Wilner D, Kawabe R (2013) The rotating outflow, envelope, and disk of the class-0/I Protostar [BHB2007]#11 in the pipe nebula. ApJ 771(2):128
9. Hogerheijde MR, van Dishoeck EF, Blake GA, van Langevelde HJ (1998) Envelope structure on 700 AU scales and the molecular outflows of low-mass young stellar objects. ApJ 502(1):315–336
10. Launhardt R, Pavlyuchenkov Ya., Gueth F, Chen X, Dutrey A, Guilloteau S, Henning Th, Piétu V, Schreyer K, Semenov D (2009) Rotating molecular outflows: the young T Tauri star in CB 26. A&A 494(1):147–156
11. Lee CF, Hirano N, Palau A, Ho PT, Bourke TL, Zhang Q, Shang H (2009) Rotation and outflow motions in the very low-mass class 0 protostellar system HH 211 at subarcsecond resolution. ApJ 699(2):1584–1594
12. Lee CF, Ho PT, Bourke TL, Hirano N, Shang H, Zhang Q (2008) SiO shocks of the protostellar jet HH 212: a search for jet rotation. ApJ 685(2):1026–1032
13. Lee CF, Ho PT, Palau A, Hirano N, Bourke TL, Shang H, Zhang Q (2007) Submillimeter arcsecond-resolution mapping of the highly collimated protostellar jet HH 211. ApJ 670(2):1188–1197
14. Lee C-F, Mundy LG, Reipurth B, Ostriker EC, Stone JM (2000) CO outflows from young stars: confronting the jet and wind models. ApJ 542(2):925–945
15. Loinard L, Rodríguez LF, D'Alessio P, Wilner DJ, Ho PT (2002) Orbital proper motions in the protobinary system L1527/IRAS 04368+2557? ApJ 581(2):L109–L113
16. Lumbreras AM, Zapata LA (2014) SMA submillimeter observations of HL tau: revealing a compact molecular outflow. AJ, 147(4):72
17. Machida MN, Hosokawa T (2013) Evolution of protostellar outflow around low-mass protostar. MNRAS 431(2):1719–1744

18. Mŭller HS, Schlŏder F, Stutzki J, Winnewisser G (2005) The cologne database for molecular spectroscopy, CDMS: a useful tool for astronomers and spectroscopists. J Molecular Struct 742(1–3):215–227
19. Ohashi N, Hayashi M, Ho PT, Momose M (1997) Interferometric imaging of IRAS 04368+2557 in the L1527 molecular cloud core: a dynamically infalling envelope with rotation. ApJ 475(1):211–223
20. Ohashi N, Saigo K, Aso Y, Aikawa Y, Koyamatsu S, Machida MN, Saito M, Takahashi SZ, Takakuwa S, Tomida K, Tomisaka K (2014) Formation of a Keplerian disk in the infalling envelope around L1527 IRS: transformation from infalling motions to Kepler motions. ApJ 796(2):131
21. Oya Y, Sakai N, Sakai T, Watanabe Y, Hirota T, Lindberg JE, Bisschop SE, Jørgensen JK, van Dishoeck EF, Yamamoto S (2014) A substellar-mass protostar and its outflow of IRAS 15398–3359 revealed by subarcsecond-resolution observations of H_2CO and CCH. ApJ 795(2):152
22. Pech G, Zapata LA, Loinard L, Rodríguez LF (2012) A rotating molecular jet from a perseus protostar. ApJ 751(1):78
23. Sakai N, Oya Y, Sakai T, Watanabe Y, Hirota T, Ceccarelli C, Kahane C, Lopez-Sepulcre A, Lefloch B, Vastel C, Bottinelli S, Caux E, Coutens A, Aikawa Y, Takakuwa S, Ohashi N, Yen H-W, Yamamoto S (2014) A chemical view of protostellar-disk formation in L1527. ApJ 791(2):L38
24. Sakai N, Sakai T, Hirota T, Watanabe Y, Ceccarelli C, Kahane C, Bottinelli S, Caux E, Demyk K, Vastel C, Coutens A, Taquet V, Ohashi N, Takakuwa S, Yen H-W, Aikawa Y, Yamamoto S (2014) Change in the chemical composition of infalling gas forming a disk around a protostar. Nature 507(7490):78–80
25. Sakai N, Sakai T, Hirota T, Yamamoto S (2008) Abundant carbon-chain molecules toward the low-mass protostar IRAS 04368+2557 in L1527. ApJ 672(1):371–381
26. Sakai N, Sakai T, Hirota T, Yamamoto S (2010) Distributions of carbon-chain molecules in L1527. ApJ 722(2):1633–1643
27. Sakai N, Yamamoto S (2013) Warm carbon-chain chemistry. Chem Rev 113(12):8981–9015
28. Satoko T, Ho PT (2012) The discovery of the youngest molecular outflow associated with an intermediate-mass protostellar core, MMS-6/OMC-3. ApJ 745(1):L10
29. Takahashi S, Ohashi N, Bourke TL (2013) Direct imaging of a compact molecular outflow from a very low luminosity object: L1521F-IRS. ApJ 774(1):20
30. Tobin JJ, Hartmann L, Calvet N, D'Alessio P (2008) Constraining the envelope structure of L1527 IRS: infrared scattered light modeling. ApJ 679(2):1364–1384
31. Tobin JJ, Hartmann L, Chiang HF, Wilner DJ, Looney LW, Loinard L, Calvet N, D'Alessio P (2012) A 0.2-solar-mass protostar with a Keplerian disk in the very young L1527 IRS system. Nature 492(7427):83–85
32. Tobin JJ, Hartmann L, Chiang H-F, Wilner DJ, Looney LW, Loinard L, Calvet N, D'Alessio P (2013) Modeling the resolved disk around the class 0 Protostar L1527. ApJ 771(1):48
33. Tobin JJ, Hartmann L, Loinard L (2010) The inner envelope and disk of L1527 revealed: gemini L'-band-scattered light imaging. ApJ 722(1):L12–L17
34. Tomisaka Kohji (2002) Collapse of rotating magnetized molecular cloud cores and mass outflows. ApJ 575(1):306–326
35. Torres RM, Loinard L, Mioduszewski AJ, Rodríguez LF (2007) VLBA determination of the distance to nearby star-forming regions. II. hubble 4 and HDE 283572 in taurus. ApJ 671(2):1813–1819
36. Van der Tak FF, Black JH, Schŏdier FL, Jansen DJ, van Dishoeck EF (2007) A computer program for fast non-LTE analysis of interstellar line spectra. With diagnostic plots to interpret observed line intensity ratios. A&A 468(2):627–635
37. van Dishoeck EF, Blake GA, Jansen DJ, Groesbeck TD (1995) Molecular abundances and low-mass star formation. II. Organic and deuterated species toward IRAS 16293-2422. ApJ 447:760
38. Velusamy T, Langer WD, Thompson T (2014) HiRes deconvolved spitzer images of 89 protostellar jets and outflows: new data on the evolution of the outflow morphology. ApJ 783(1):6

39. Watanabe Y, Sakai N, Lindberg JE, Jørgensen JK, Bisschop SE, Yamamoto S (2012) An unbiased spectral line survey toward R CrA IRS7B in the 345 GHz window with ASTE. ApJ 745(2):126
40. Xue Y, Suto Y, Taruya A, Hirano T, Fujii Y, Masuda K (2014) Tidal evolution of the spin-orbit angle in exoplanetary systems. ApJ 784(1):66
41. Yeh SC, Hirano N, Bourke TL, Ho PT, Lee CF, Ohashi N, Takakuwa S (2008) The CO molecular outflows of IRAS 16293-2422 probed by the submillimeter array. ApJ 675(1):454–463
42. Yen HW, Takakuwa S, Ohashi N, Ho PT (2013) Unveiling the evolutionary sequence from infalling envelopes to keplerian disks around low-mass protostars. ApJ 772(1):22
43. Zapata LA, Schmid-Burgk J, Muders D, Schilke P, Menten K, Guesten R (2010) A rotating molecular jet in Orion. A&A 510:A2
44. Zapata LA, Arce Héctor G, Brassfield E, Palau A, Patel N, Pineda JE (2014) A spider-like outflow in Barnard 5 - IRS 1: the transition from a collimated jet to a wide-angle outflow? MNRAS 441(4):3696–3702

Chapter 5
IRAS 15398–3359

5.1 Introduction

In L1527, it has been demonstrated that the kinematic structures of the infalling-rotating envelope and the outflow cavity wall are explained by the simple models (Chap. 4). Especially, the discovery of the centrifugal barrier in an infalling-rotating envelope and the chemical change across the centrifugal barrier provide us with a completely new insight in disk formation study. It is now required to confirm whether such kinematic and chemical structures are common in low-mass protostellar sources. In this chapter, a similar analysis is conducted in another WCCC source IRAS 15398−3359. While the structure of the outflow in L1527 has been analyzed based on the analysis of its envelope, the results from the clear outflow components in IRAS 15398−3359 is utilized to investigate the envelope structure.

It is still unknown when and how disks are formed in the star formation process. This is because that overwhelming emission from protostellar envelopes and outflows generally make it difficult to identify possible disks in Class 0 sources. Even absence of such a disk down to 45 au is claimed for the Class 0 protostar NGC 1333 IRAS 2A [3, 13]. Since possible disk structures in Class 0 sources are expected to be compact, high spatial-resolution and high sensitivity observations are indispensable to detect them.

IRAS 15398−3359 is a low-mass Class 0 protostar in the Lupus 1 molecular cloud ($d \sim 155$ pc [11]). A molecular outflow was detected by single-dish observations of CO emission [27, 30]. The CO ($J = 3 - 2$) observation with the James Clerk Maxwell Telescope ($\theta_{\mathrm{HPBW}} \sim 15''$) showed that the red- and blue-shifted lobes of this source were relatively overlapped with each other, and thus a pole-on geometry was suggested [30]. [12] observed H_2CO and CS lines toward this source to search for a sign of an infall motion without success. On the other hand, [9] reported the presence of an inverse P-Cygni profile of the H_2O ($1_{1,0} - 1_{0,1}$) at 557 GHz, indicating an infall

This chapter has been published in Oya et al., 2014, ApJ, 795, 152. © AAS. Reproduced with permission.

Y. Oya, *A Few Tens au Scale Physical and Chemical Structures Around Young Low-Mass Protostars*, Springer Theses, https://doi.org/10.1007/978-981-19-1708-0_5

motion of the envelope on a scale of about 10^4 au. [7] detected a ring structure in the $H^{13}CO^+$ ($J = 4 - 3$) line at a (150 – 200) au scale. They proposed that a recent accretion burst in this source enhanced the luminosity leading the H_2O evaporation, which would destruct HCO^+ in the vicinity of the protostar and cause the observed ring structure.

IRAS 15398−3359 shows peculiar chemical features. Various carbon-chain molecules, such as CCH, C_4H, and CH_3CCH are detected toward this source, which is characteristic to so-called warm carbon-chain chemistry (WCCC) [22, 24, 25]. As demonstrated in Chaps. 3 and 4, the kinematic structure of the envelope in another WCCC source, L1527, is well reproduced by the infalling-rotating envelope model. Because of the chemical resemblance of IRAS 15398−3359 to L1527, it is interesting to compare the physical structures of their envelopes. A deep insight into envelope structures of Class 0 protostellar sources is essential to explore when and how rotationally-supported disks are formed around protostars. With these motivations, several molecular lines are observed with ALMA toward IRAS 15398−3359.

5.2 Observations

ALMA observations toward IRAS 15398−3359 were carried out on December 31st, 2012 during its Cycle 0 operations. Spectral lines of CCH and H_2CO were observed with the Band 7 receiver at frequencies of (349 – 364) GHz, as listed in Table 5.1. These observations were carried out with twenty-five antennas with the baseline length ranging from 13 to 338 m. The field center of the observations was set to be $(\alpha_{2000}, \delta_{2000}) = (15^h43^m02\overset{s}{.}3, -34°09'07\overset{''}{.}5)$. The system temperature was typically from 120 to 300 K. The frequency resolution of the backend correlator was tuned to be 122 kHz and bandwidth of 469 MHz, corresponding to the velocity resolution of 0.1 km s^{-1} at 366 GHz. The phase calibrations were carried out on J1517−243 for every 12 min. J1256−057 and J1924−292 were used for the bandpass calibrations. The absolute flux density scale was derived from Mars and Titan. The data calibration was performed in the antenna-based manner where typical uncertainties are expected to be less than 10%. The primary beam (half-power beam width) is 17″. CLEAN algorithm was employed to obtain the continuum and line images. The continuum image was obtained by averaging line-free channels. After subtracting the continuum emission directly from the visibilities, the line images were obtained. The total on-source time was 27 minutes and 21 minutes for the H_2CO lines and the CCH lines, respectively. The synthesized-beam size is $0\overset{''}{.}57 \times 0\overset{''}{.}42$ (P.A. = 49°) and $0\overset{''}{.}60 \times 0\overset{''}{.}44$ (P.A. = 46°) for the continuum image and the H_2CO image, respectively. The rms noise levels are 1 and 10 mJy $beam^{-1}$ for the continuum and the H_2CO emission, respectively. The continuum emission has its peak intensity at: $(\alpha_{2000}, \delta_{2000}) = (15^h43^m02\overset{s}{.}24, -34°09'06\overset{''}{.}7)$. The CCH data were combined with those taken in another ALMA observation program (2011.0.00628.S; PI: Jes Jørgensen), which was carried out with 15 – 16 antennas. The signal-to-noise ratio (S/N) was improved from 13.7 to 24.5 by this combination. The synthesized beam

Table 5.1 Observed Line Parameters[a]

Molecule	Transition	Frequency (GHz)	$E_u k_B^{-1}$ (K)	$S\mu^{2b}$ (D^2)
H_2CO	$5_{0,5} - 4_{0,4}$	362.7360480	52	27.168
	$5_{1,5} - 4_{1,4}$	351.7686450	62	26.096
	$5_{2,4} - 4_{2,3}$	363.9458940	100	22.834
CCH	$N = 4 - 3, J = 7/2 - 5/2, F = 3 - 2$	349.4006712	42	1.6942
	$N = 4 - 3, J = 7/2 - 5/2, F = 4 - 3$	349.3992756	42	2.2712

[a]Taken from CDMS [15]

[b] Nuclear spin degeneracy is not included.

of the combined CCH image is 0.″70 × 0.″46 (P.A. = 72°). Its rms noise level is 15 mJy beam^{-1}. No other significant line features were detected in these observations, except for the line emissions reported by [7].

5.3 Results

5.3.1 *Overall Distribution of H_2CO and CCH*

Figure 5.1a shows the moment 0 (integrated intensity) map of the H_2CO ($5_{1,5} - 4_{1,4}$) line. A well-collimated outflow is detected prominently, symmetrically extending along the northeast-southwest axis from the protostellar position. The wall of this outflow has a straight structure, as like as the outflow of HH46 [1]. Figure 5.1b shows the velocity map (moment 1 map), showing that the northeastern and southwestern outflow lobes are red-shifted and blue-shifted, respectively. In these observations, the emission extending over 12″ scale or larger would be affected by the resolving-out effect due to the lack of short baseline data. Although the size of the detected outflow is as compact as 8″ for each lobe, the emission from the outer part may be resolved out. As well, it may be weak due to insufficient excitation conditions for the observed molecular lines. At a distance of 8″ from the protostellar position, the apparent width of the outflow is about 4″. Figure 5.1c shows the integrated intensity map (moment 0 map) of the high-excitation H_2CO ($5_{2,4} - 4_{2,3}$) line. The emitting region of the H_2CO ($5_{2,4} - 4_{2,3}$) line is essentially similar to that of the H_2CO ($5_{1,5} - 4_{1,4}$) line, although the S/N ratio is rather poor for the former line. In these maps, a bright knot is detected in the red-shifted lobe ('Clump A' in Fig. 5.1a). This component could be a shocked region caused by an impact of the outflow with dense clumps in the surrounding cloud.

As well as the outflow, a single-peaked compact component is also recognized in the vicinity of the protostar in Fig. 5.1d (a blow-up of the central part of Fig. 5.1a).

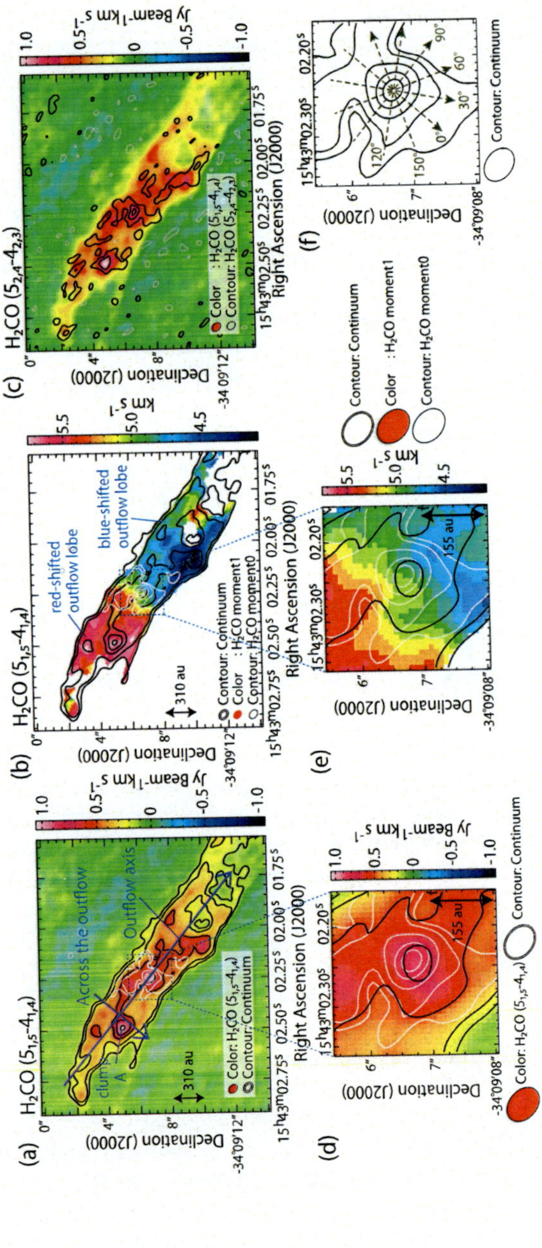

Fig. 5.1 **a** Integrated intensity map (moment 0) of the H_2CO $(5_{1,5} - 4_{1,4})$ line (color; black contours). White contours show the 0.8 mm continuum emission. The PV diagrams in Figs. 5.5a, b are prepared along the blue arrows. **b** Velocity map (moment 1 map) of the H_2CO $(5_{1,5} - 4_{1,4})$ line. Black and white contours are the same as those in panel (**a**). **c** Integrated intensity map of the H_2CO $(5_{2,4} - 4_{2,3}$; contours) line. The color map is the same as that in panel (**a**). A few bright spots are seen in the both lines of H_2CO. **d** A blow-up of the integrated intensity map of the H_2CO $(5_{1,5} - 4_{1,4})$ line (panel **a**) in the vicinity of the protostar. **e** A blow-up of the velocity map of the H_2CO $(5_{1,5} - 4_{1,4})$ line (panel **b**) in the vicinity of the protostar. **f** A blow-up of the 0.8 mm continuum map in the vicinity of the protostar. The PV diagrams in Figs. 5.7, 5.9, and 5.10 are prepared along the dashed arrows. The arrow labeled as "0°" has the position angle (P.A.) of 140°, and others are taken for P.A. of every 30°.

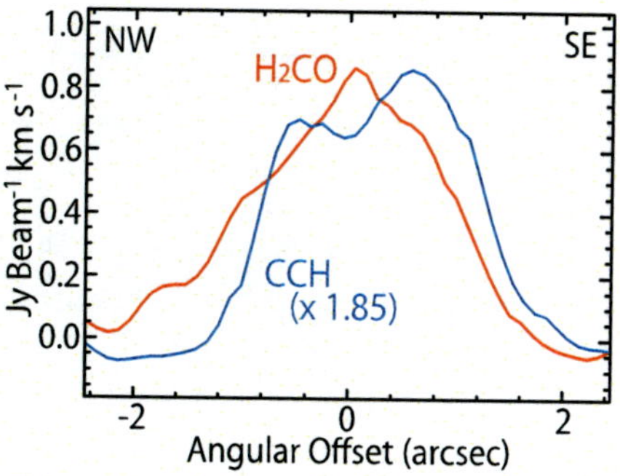

Fig. 5.2 Intensity profiles of the H_2CO ($5_{1,5}-4_{1,4}$) line and the CCH ($N=4-3$, $J=7/2-5/2$, $F=4-3$ and $3-2$) lines along the line perpendicular to the outflow axis ("0°" shown in Fig. 5.1f)

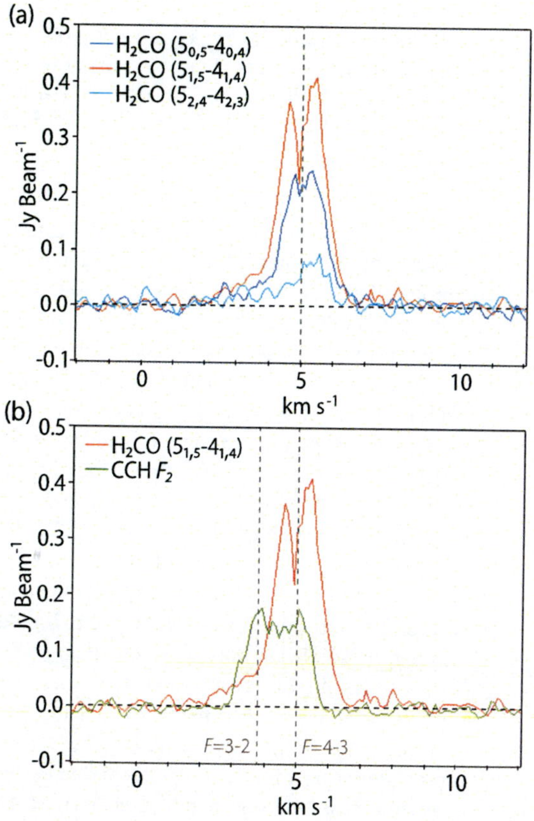

Fig. 5.3 Spectra of the H_2CO ($5_{0,5}-4_{0,4}$, $5_{1,5}-4_{1,4}$ and $5_{2,4}-4_{2,3}$) lines and the CCH ($N=4-3$, $J=7/2-5/2$, $F=4-3$ and $3-2$) lines prepared toward the protostellar position. In panel (**b**), the systemic velocities for the two hyperfine components of CCH are represented by the two vertical dashed lines

Figure 5.1e represents a blow-up of the velocity map (Fig. 5.1b). The approximate extent of the central component is estimated to be about 2″ (310 au) in diameter based on the spatial profile of the line intensity prepared along the line perpendicular to the outflow axis (Fig. 5.2). In Fig. 5.3, the spectral line profiles of the H_2CO and CCH

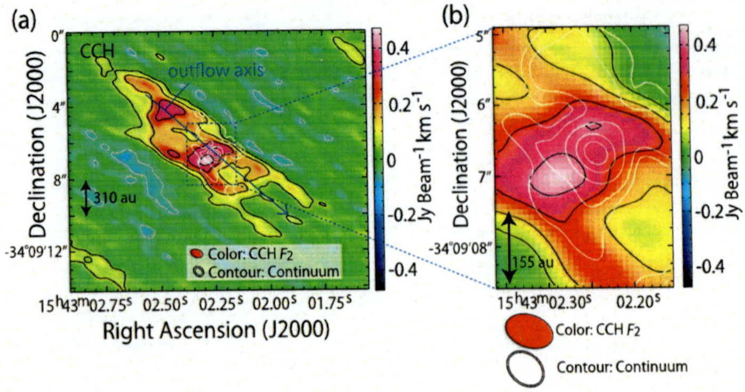

Fig. 5.4 **a** Integrated intensity map of the CCH ($N = 4 - 3$, $J = 7/2 - 5/2$, $F = 4 - 3$ and $3 - 2$; color, black contour) lines. White contours represent the 0.8 mm continuum map as the same as those in Fig. 5.1a. The PV diagram in Fig. 5.6 is prepared along the blue arrow parallel to the outflow axis. **b** A blow-up of panel (**a**) in the vicinity of the protostar. Contours are as the same as those in panel (**a**)

lines are shown, which are prepared at the protostellar position and averaged over the beam size. The line width of the H_2CO line is as narrow as 2 km s^{-1} even toward the protostellar position. By using the RADEX program [29] to fit the intensity of the two temperature sensitive lines of para-H_2CO ($5_{0,5} - 4_{0,4}$, $5_{2,4} - 4_{2,3}$), the column density of H_2CO and its kinetic temperature toward the protostellar position are estimated to be 3×10^{13} cm^{-2} and 36 − 38 K, respectively, assuming the H_2 density of $(1 \times 10^7 - 1 \times 10^8)$ cm^{-3} and the line width of 1.8 km s^{-1} for the H_2CO line. As well, the ortho/para ratio is estimated to be 2.8 based on the intensity of the ortho-H_2CO ($5_{1,5} - 4_{1,4}$) line. The optical depths for these lines are 0.27 ($5_{0,5} - 4_{0,4}$), 0.06 ($5_{2,4} - 4_{2,3}$), 0.53 ($5_{1,5} - 4_{1,4}$), and therefore, the lines are not opaque.

The integrated intensity map of the CCH ($N = 4 - 3$, $J = 7/2 - 5/2$, $F = 4 - 3$ and $3 - 2$) lines is shown in Fig. 5.4a, and its blow-up is in Fig. 5.4b. The outflow cavity is prominently traced as well as the centrally concentrated component is also seen. The emitting region of the CCH line around the protostar is more extended along the southeast-northwest axis than that of the H_2CO line. The CCH emission shows a slight dip toward the protostellar position, as shown in its intensity profile prepared along the line perpendicular to the outflow axis (Fig. 5.2). This feature is consistent with the previous report [7]. The spectrum of the CCH lines taken at the protostellar position is complicated due to the two hyperfine components (Fig. 5.3). Nevertheless, the line width of each hyperfine component of CCH is found to be 2 km s^{-1} or less, as like as that of the H_2CO line.

5.3.2 Outflow

First, the outflow feature detected in the H_2CO $(5_{1,5} - 4_{1,4})$ line is analyzed. Figure 5.5a shows the position-velocity (PV) diagram of the H_2CO $(5_{1,5} - 4_{1,4})$ line. Its position axis is shown by a blue arrow in Fig. 5.1a, which is centered at the protostellar position and prepared along the outflow axis. The observed outflow structure extends along the northeast-southwest axis from the protostellar position, where the northeastern and southwestern parts are red-shifted and blue-shifted, respectively. Since the red-shifted and blue-shifted outflow lobes show little overlap with each other near the protostar, the outflow likely blows almost perpendicular to the plane of the sky, which indicates that the disk/envelope geometry has a nearly edge-on configuration. As often observed for outflow cavities (e.g. [1, 10]), the highest velocity shift in Fig. 5.5a is found to increase linearly as the distance from the protostar. An intense knot is detected in the red-shifted component, corresponding to Clump A in the integrated intensity map (Fig. 5.1a). As well as the high velocity-shift components, other components are found near the systemic velocity (~ 5 km s^{-1}; [22]). The low velocity-shift components are slightly blue-shifted and red-shifted on the northeastern and southwestern side of the protostar. This trend is the opposite case to the high velocity-shift components.

Figure 5.5b shows the PV diagram of the H_2CO $(5_{1,5} - 4_{1,4})$ line in the red-shifted outflow lobe, whose position axis is prepared to be perpendicular to the outflow axis as shown in Fig. 5.1a. This diagram shows an elliptic feature. As well, a knot-like emission is also detected in the higher-velocity range. This emission corresponds to Clump A found in the integrated intensity map (Fig. 5.1a). The velocity structure shown in the PV diagram likely suggests the expanding motion of the gas in the outflow cavity wall.

In this study, the standard model of an outflow cavity from [10] is employed to analyze the observed geometrical and kinematical structures of the outflow, which

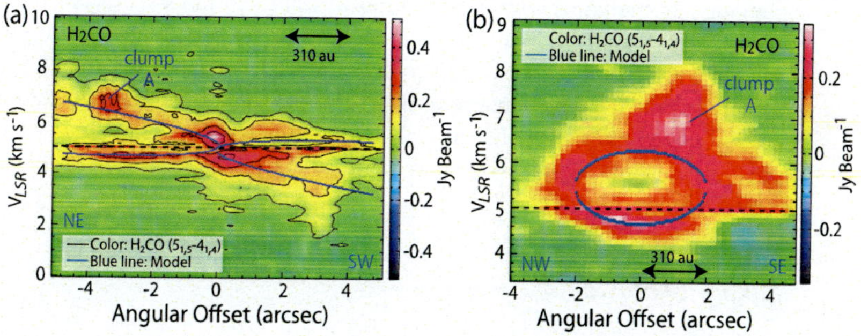

Fig. 5.5 PV diagrams of the H_2CO $(5_{1,5} - 4_{1,4})$ line **a** along the outflow axis and **b** the line perpendicular to it. Their position axes are represented by the blue arrows in Fig. 5.1a. Blue lines show the best outflow model with an inclination angle of 70° (0° for a pole-on configuration). The systemic velocity is represented by the black horizontal line in each panel

is employed for the outflow analysis in L1527 (Chap. 4). In this model, a parabolic shape is assumed for the outflow cavity wall, and the velocity of the gas on the wall is assumed to be proportional to the distance to the protostar;

$$z = C_{\mathrm{as}} R^2, \quad v_R = v_{\mathrm{as}} \frac{R}{R_0}, \quad v_z = v_{\mathrm{as}} \frac{z}{z_0}, \tag{5.1}$$

where z denotes the distance to the protostar along the outflow axis, and R the radial size of the cavity wall perpendicular to z-axis (see Chap. 3). Both z_0 and R_0 are normalization constants set to be 1″. C_{as} and v_{as} are free parameters. In this model, the intensity of the emission is assumed to be proportional to the column density, where the molecular density is simply assumed to be constant. These assumptions may not be realistic. However, they do not matter the following analysis, because this study focuses on the velocity structure but not on the intensity profile. The best results are obtained with an inclination angle of 70° (0° for a pole-on configuration), as represented by the blue lines in Fig. 5.5. With an inclination angle less than 60° or higher than 80°, the outflow model does not reproduce the observed kinematic structure well with any values of C_{as} and v_{as}. Hence, the inclination angle is determined to be 70° ± 10°, where the error value is the estimated limit based on the above results. The derived parameters are C_{as} of 0.8 arcsec^{-1} and v_{as} of 0.38 km s^{-1} for an inclination angle of 70°. As mentioned in Sect. 5.1, [30] reported an inclination angle of 15°, a nearly pole-on geometry. This discrepancy seems to originate from the limited spatial resolution of their data. Another possibility is that the outflow direction at a small scale is different from that on larger scales (e.g. [8, 14, 20, 32]), as discussed in the analysis of L1527 (Cha. 4). On the other hand, Atacama Pathfinder EXperiment observations of the CO ($J = 6 - 5$) line indicate an inclination angle of 70° [31], which is consistent with the above result.

Figure 5.6 shows the PV diagram of the CCH ($N = 4 - 3$, $J = 7/2 - 5/2$, $F = 4 - 3$ and $3 - 2$) line prepared along the outflow axis through the protostellar position. The appearance of the PV diagram is complicated due to the two hyperfine components blending with a separation of only 1.2 km s^{-1}. Compared with the H_2CO case, the low velocity-shift components of CCH is relatively bright in comparison with its high velocity-shift components. This suggests that the CCH emission from the outflow component primarily traces the compressed ambient gas around the outflow cavity rather than the entrained outflowing gas. Gas-phase reactions in dense photodissociation region layers may form CCH (e.g. [5, 6, 26]).

5.3.3 *Protostellar Envelope*

A centrally concentrated component is detected in both the H_2CO and CCH emission (Figs. 5.1d, 5.4b). The distributions of $C^{34}S$, $C^{17}O$, and CH_3OH also have such a component [7]. In this section, the kinematic structure of this component is investigated. A blow-up of the velocity map of the H_2CO line (Fig. 5.1e) is dominated by

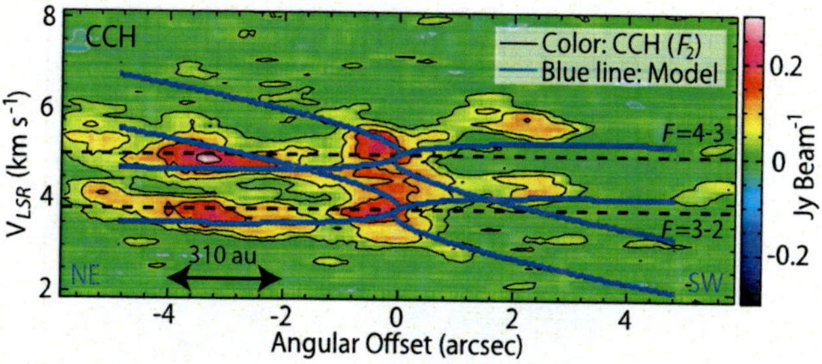

Fig. 5.6 PV diagram of the CCH ($N = 4 - 3$, $J = 7/2 - 5/2$, $F = 4 - 3$ and $3 - 2$) lines. The position axis is taken along the outflow axis, which is shown by a blue arrow in Fig. 5.4a. Blue lines in the diagram represent the outflow model results with an inclination angle of 70°, which is the best model obtained in the H_2CO analysis (Sect. 5.3.2). The systemic velocities for the two hyperfine components of CCH (5.0 km s^{-1}, 3.8 km s^{-1}) are shown by the black dashed lines

the overwhelming emission from the outflow components. Nevertheless, a slightly skewed feature is marginally recognized around the continuum peak position. In order to reveal the gas motion in the vicinity of the protostar more carefully, PV diagrams around the protostar are prepared (Fig. 5.7).

The PV diagrams of the H_2CO ($5_{1,5} - 4_{1,4}$) line are shown in Fig. 5.7a, b. They are prepared along the lines shown in Fig. 5.1f, centered at the protostellar position; Fig. 5.7a is along the line perpendicular to the outflow axis ("0°"), while Fig. 5.7b along the outflow axis ("90°"). Although a rotation signature is not obvious in the centrally concentrated component in Fig. 5.7a, there is a marginal trend that the intensity peaks in the red-shifted and the blue-shifted velocity ranges are on the southeastern and northwestern sides of the protostellar position, respectively. In fact, this rotation motion has been later confirmed [17].

Meanwhile, Fig. 5.7b shows two intensity peaks in the vicinity of the protostar; one on the northeastern side of the protostar is red-shifted, the other on the south-western side is blue-shifted. This systematic velocity gradient is the same case of that found in the outflow (Fig. 5.5a). However, the modest velocity gradient near the protostar is difficult to attribute to the outflow component, according to the outflow model result obtained in Sect. 5.3.2. Based on the outflow direction and its inclination angle, the outflow and disk/envelope structure of IRAS 15398−3359 is schematically illustrated as shown in Fig. 5.8. Therefore, the velocity gradient found in Fig. 5.7b is likely a signature of an infall motion of the envelope gas rather than an outflow motion.

An infall/rotation signature is not clearly seen in the PV diagrams for the CCH ($N = 4 - 3$, $J = 7/2 - 5/2$, $F = 4 - 3$ and $3 - 2$) lines (Fig. 5.7c, d), in contrast to the L1527 case (Chapters 3, 4 [21, 23]). In L1527, whose protostellar mass and inclination angle are derived to be 0.18 $M_\odot$ and 85°, respectively, the observed velocity shift of the CCH line is 1.8 km s^{-1} at a radius of 100 au from the protostar. On the contrary, the velocity shifts are as small as 1 km s^{-1} and 0.7 km s^{-1} in the H_2CO and CCH lines, respectively, in IRAS 15398−3359 (Fig. 5.7), although a nearly edge-

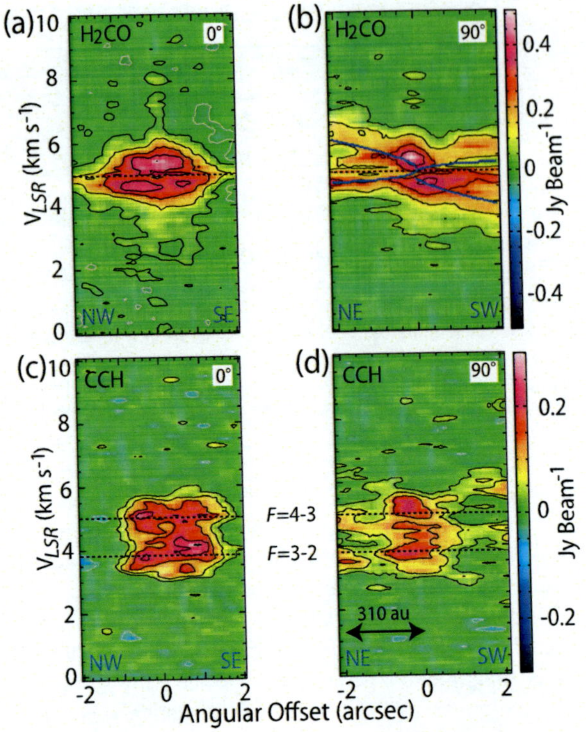

Fig. 5.7 PV diagrams of the H_2CO ($5_{1,5} - 4_{1,4}$; a, b) and CCH ($N = 4 - 3$, $J = 7/2 - 5/2$, $F = 4 - 3$ and $3 - 2$; c, d) lines. Their position axes are shown in Fig. 5.1f; those for panels (**a**) and (**c**) are taken along the line perpendicular to the outflow axis ("0°"), while those for (**b**) and (**d**) are along the outflow axis ("90°"). Panels (**b**) and (**d**) are blow-ups of Figs. 5.5a and 5.6, respectively. Blue lines in panel (**b**) represent the best model for the outflow component (Sect. 5.3.2)

on configuration (70°; 0° for a face-on configuration) is expected according to the outflow analysis (Sect. 5.3.2). Although IRAS 15398−3359 is similar to L1527 in its large-scale ($\sim$ a few 1000 au) chemical composition [22], infall/rotation motions are not very clear contrarily.

In these observations, the structure down to 78 au (0.″5) are resolved around the protostar. Nevertheless, the observed line widths are quite narrow. Considering the nearly edge-on configuration, it is suggested that the Doppler shift due to infall/rotation motions around the protostar is small. A small velocity-shift could be explained if the observed dense gas were not associated with the protostar, that is, if it were mostly present in the outflow component. However, this is not likely the case because the modeled outflow motion does not well explain the velocity gradient traced by the H_2CO line (Fig. 5.7b), as mentioned above. Hence, the small velocity-shift likely implies a low protostellar mass. A rough upper limit for the protostellar mass is estimated based on the maximum value of the observed velocity shifts. Under the conservation law of energy, the protostellar mass (M) is represented in terms of the infall velocity (v_{infall}) and the rotation velocity (v_{rotation}) as:

$$M = \frac{r}{2G}\left(v_{\mathrm{infall}}^2 + v_{\mathrm{rotation}}^2\right), \tag{5.2}$$

where r denotes the distance to the protostar. By using this relation, the protostellar mass is roughly estimated to be smaller than 0.09 $M_\odot$ with a maximum velocity less than 1 km s^{-1} at 0.″5 (78 au) from the protostar. Here, both the infall and rotation velocities are assumed to be the observed maximum value of 1 km s^{-1} in Fig. 5.7a as a robust case. If there is no rotation motion, i.e. the free fall case, the upper limit for the protostellar mass is obtained to be 0.04 $M_\odot$. This rough estimate is indeed consistent with the later report with a further observation [17].

5.3.4 Comparison with an Envelope Model

A ballistic model of an infalling-rotating envelope (Fig. 5.8) is employed to investigate the observational results of the c-C_3H_2 and CCH lines toward L1527 (Chaps. 3

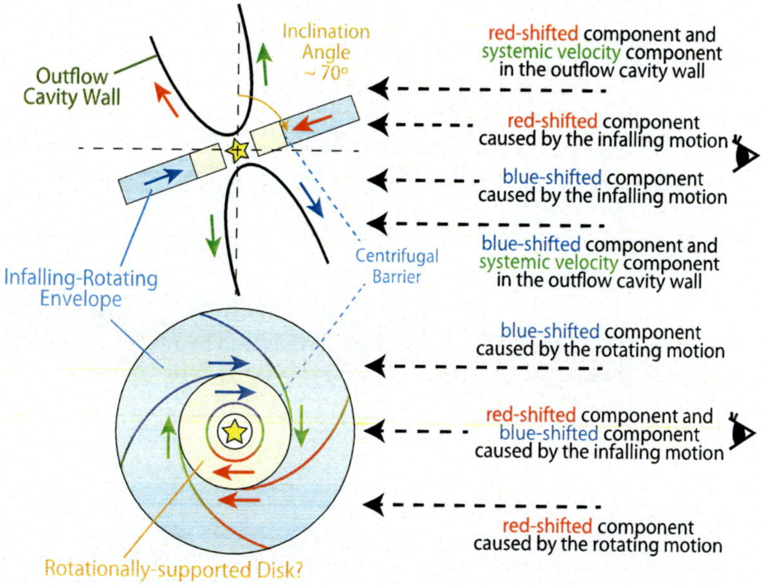

Fig. 5.8 Schematic illustrations of the outflow and disk/envelope system. An observer sits on the right hand side, whose line of sight is shown by the black dashed arrows. The disk/envelope system has an almost edge-on geometry with an inclination angle of 70° (0° for a face-on configuration)

and 4). This source has similar chemical characteristics to IRAS 15398−3359. In spite of the simplicity of this model, the basic features of the PV diagrams of the c-C_3H_2 and CCH lines toward L1527 are well reproduced. Therefore, it is worth applying the same model to the kinematic structure traced by the H_2CO and CCH lines in IRAS 15398−3359. The CCH emission observed in L1527 only traces the infalling-rotating envelope, and thus, it is preferable to apply the infalling-rotating envelope model for the CCH line. However, the PV diagrams of different transitions of CCH observed toward IRAS 15398−3359 are complicated due to the hyperfine structure, and their signal-to-noise ratios are insufficient for critical comparison. In contrast, the H_2CO ($5_{1,5} - 4_{1,4}$) emission is bright enough in IRAS 15398−3359. Although H_2CO possibly resides in the inner disk-like structure in addition to the infalling-rotating envelope, the contribution of the envelope seems significant in L1527 [21]. Hence, the model results are compared with the H_2CO ($5_{1,5} - 4_{1,4}$) line observed in IRAS 15398−3359.

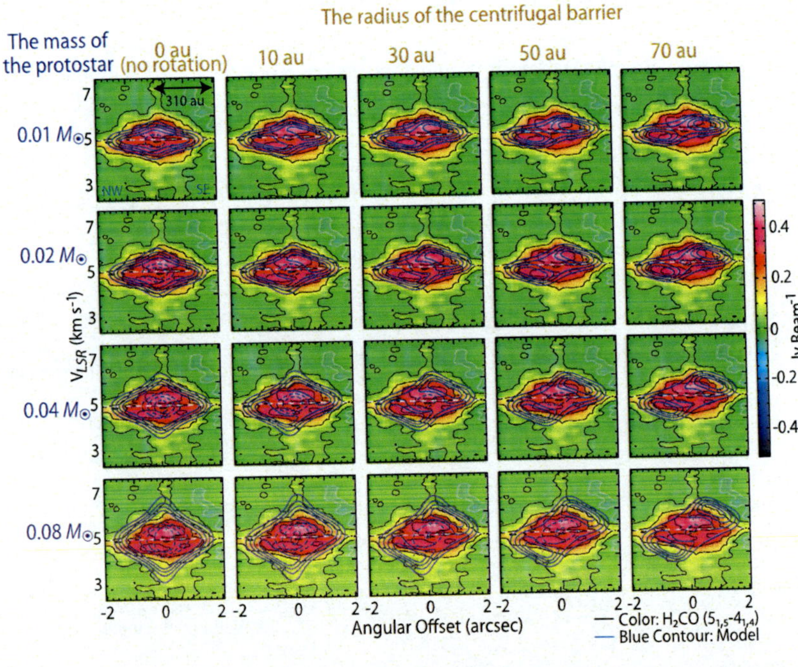

Fig. 5.9 PV diagrams of the H_2CO ($5_{1,5} - 4_{1,4}$) line prepared along the line perpendicular to the outflow axis. Their position axes are represented by the arrow labeled as "0°" in Fig. 5.1f. Blue contours represent 20 model results of the infalling-rotating envelope with an inclination angle of 70°. The protostellar mass and the radius of the centrifugal barrier are varied as free parameters (see Chap. 3). Contour levels are every 20% of the peak intensity in each panel. Dashed contours around the central position represent the intensity dip

Figure 5.9 shows the PV diagrams of the H_2CO ($5_{1,5} - 4_{1,4}$) line along the line perpendicular to the outflow axis (the arrow labeled as "0°" in Fig. 5.1f), that is, along the mid-plane of the envelope. The blue contours represent the results of the infalling-rotating envelope model. In this model, the intensity is simply assumed to be proportional to the column density. The molecular density is assumed to be proportional to $r^{-1.5}$, where r denotes the distance to the protostar. The assumption of the optically thin condition seems to be reasonable, because the optical depth is evaluated to be less than 0.3 (Sect. 5.3.1). The excitation effect and the effect of the temperature gradient are not taken into account, neither. However, these simplifications may not affect the following analysis, because this analysis only focuses on the velocity structure but not on the intensity profile. In contrast to the L1527 case (Chap. 4), the radius of the centrifugal barrier cannot be directly determined from the PV diagram because of the absence of an obvious rotation signature. Nevertheless, its upper limit can be roughly estimated based on the CCH distribution. If CCH resides only in the infalling-rotating envelope as in the case of L1527 and the radius of the centrifugal barrier is larger than the beam size, a hole in the CCH distribution toward the protostellar position is expected to be resolved. Although an intensity dip toward the protostar is marginally seen in the CCH emission (Fig. 5.2), this feature is not well resolved in the present observation with the beam size of $0''.5$ (78 au). Hence, the upper limit for the centrifugal barrier is set to be 80 au. Under this constraint, Fig. 5.9 shows the simulated PV diagrams of the infalling-rotating envelope models with various sets of the protostellar mass (M) and the radius of the centrifugal barrier (r_{CB}). Among the 20 models in Fig. 5.9, ones with M of 0.02 $M_{\odot}$ and r_{CB} from 0 to 30 au are chosen as the best models by eye. The value of M is consistent with its upper limit obtained in Sect. 5.3.3. Thus, the low protostellar mass of this source is confirmed with this model analysis. The PV diagrams of the model and observation along the different directions are also shown in Fig. 5.10; the diagrams are prepared along the six lines shown in Fig. 5.1f, and the model with M of 0.02 $M_{\odot}$ and r_{CB} of 30 au are employed as an example. The observed trends of velocity gradients around the protostar seem to be explained by this model.

5.4 Discussion

The small protostellar mass of IRAS 15398−3359 ($< 0.09\ M_{\odot}$) is essentially derived from the narrow line width detected in the inner envelope ($<$ 80 au) with an almost edge-on configuration (70°; 0° for a face-on configuration). The mass estimates may suffer from the uncertainty in the inclination angle of the disk/envelope system, since the inclination angle evaluated for the outflow structure is employed as the substitute. If the inclination angle is smaller than the employed value, i.e. nearer to the face-on configuration, the mass evaluation by the model increases. Nevertheless, the upper limit value for the protostellar mass does not change significantly, even if an inclination angle of 60° is employed. In fact, the upper limit value was confirmed

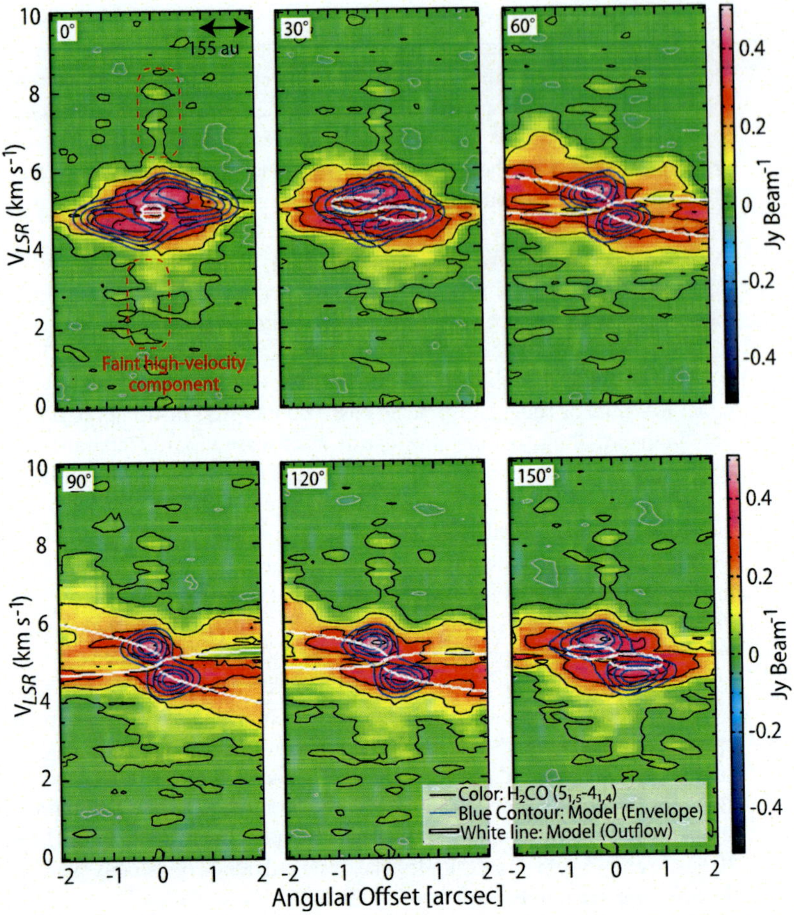

Fig. 5.10 PV diagrams of the H_2CO ($5_{1,5} - 4_{1,4}$) line prepared centered at the protostellar position. Their position axes are represented by the six arrows in Fig. 5.1f, which are taken for different position angles. Blue contours represent the results of the infalling-rotating envelope model. The parameters for the model are the inclination angle of 70°, the protostellar mass of 0.02 $M_{\odot}$, and the radius of the centrifugal barrier of 30 au. Contour levels are every 20% of the peak intensity in each panel. White lines show the outflow model obtained in Sect. 5.3.2

to be adequate later based on the further observations [17], where the protostellar mass was evaluated to be as small as 0.007 $M_{\odot}$.

The averaged accretion rate is evaluated to be less than 9.0×10^{-5} $M_{\odot}$ yr^{-1} by using the dynamical timescale of the extended outflow ($(1 - 2) \times 10^3$ yr for each lobe [31]) and the upper limit for the protostellar mass (< 0.09 $M_{\odot}$) obtained in Sect. 5.3.3. Because older parts of the outflow structure may not be detected, the evaluated dynamical timescale of the outflow should be regarded as its lower limit. Thus, the above value for the averaged accretion rate is its upper limit. Nevertheless, the estimated accretion rate seems roughly consistent with the typical value for low-mass protostars from 10^{-5} to 10^{-6} $M_\odot$ yr^{-1} (e.g. [4]), and may be higher than that for another WCCC source L1527 (10^{-6} $M_\odot$ yr^{-1} [16]). Meanwhile, the mass outflow rate is calculated to be 6.3×10^{-6} $M_\odot$ yr^{-1} and 2.2×10^{-6} $M_\odot$ yr^{-1} for the red- and blue-shifted lobes, respectively, by using their masses and dynamical timescales reported by [30]. These values are modified to be 1.7×10^{-5} $M_\odot$ yr^{-1} and 6.2×10^{-6} $M_\odot$ yr^{-1} including the correction for the inclination angle of 70°. Therefore, the averaged mass accretion rate and the mass outflow rate seem to be roughly comparable to each other.

Meanwhile, the mass accretion rate ($\dot{M}$) can be estimated by using the following equation [19]:

$$\dot{M} = \frac{LR_{\rm star}}{GM}, \tag{5.3}$$

where M, L, and $R_{\rm star}$ denote the mass, luminosity, and radius of the protostar, respectively. With the upper limit for M (< 0.09 $M_\odot$) obtained in Sect. 5.3.3, $\dot{M}$ is estimated to be larger than 1.6×10^{-6} $M_\odot$ yr^{-1} by employing L of 1.8 $L_\odot$ [7] and $R_{\rm star}$ of 2.5 $R_\odot$ (e.g. [2, 18]). This lower limit value for $\dot{M}$ is consistent with its upper limit obtained above (9.0×10^{-5} $M_\odot$ yr^{-1}). In spite of the episodic accretion in this source suggested by [7], the averaged mass accretion rate does not significantly differ from the canonical value ($10^{-5} - 10^{-6}$ $M_\odot$ yr^{-1}).

The upper limit of the protostar/envelope mass ratio is evaluated to be 0.18 by using the upper limit for the protostellar mass of 0.09 $M_\odot$ and an envelope mass of 0.5 $M_\odot$ [9]. Employing the protostellar mass of 0.02 $M_\odot$, the ratio is calculated to be 0.04. Thus, the ratio in this source seems smaller than that in L1527 (0.2 [28]). This result implies that the protostar is in its infant stage and is still growing, which was later supported with further investigation [17]. Since the specific angular momentum of the gas brought into the vicinity of the protostar is expected to be smaller at the earlier stage in the protostar evolution, it would be a natural consequence that the rotation signature is not evident in IRAS 15398−3359 in contrast to the case in L1527 (Chap. 4), a more evolved Class I source.

In Fig. 5.2, the H_2CO emission shows a centrally concentrated distribution with a single peak, whereas the CCH emission has a more flattened distribution. Therefore, the H_2CO emission is expected to trace the inner region than the CCH emission. A similar trend is also seen in L1527 [21]. It should be noted that Fig. 5.10 shows a faint high-velocity component of H_2CO ($v_{\rm lsr} < 4$ km s^{-1} or $v_{\rm lsr} > 6$ km s^{-1}) toward the protostellar position. If this faint emission is really the contribution from a rotationally-supported disk, it means that a disk structure is already formed at such a very early stage of the low-mass star-formation. With the upper limit for the protostellar mass of 0.09 $M_\odot$, the maximum velocity shift about 3 km s^{-1} in the H_2CO line corresponds to the radius of 10 au.

As demonstrated in this chapter, extensive studies of the disk/envelope structure in a Class 0 source have now become possible with ALMA, and a chemical approach is helpful for such studies. This study reveals that the disk/envelope structure can be traced with various molecular lines. Hence, the chemical evolution can be addressed even in the disk-forming stage.

References

1. Arce HG, Mardones D, Corder SA, Garay G, Noriega-Crespo A, Raga AC (2013) ALMA observations of the HH 46/47 molecular outflow. ApJ 774(1):39
2. Baraffe I, Chabrier G (2010) Effect of episodic accretion on the structure and the lithium depletion of low-mass stars and planet-hosting stars. A&A 521:A44
3. Brinch C, Jørgensen JK, Hogerheijde MR (2009) The kinematics of NGC 1333-IRAS2A - a true Class 0 protostar. A&A 502(1):199–205
4. Hartmann L, Cassen P, Kenyon SJ (1997) Disk accretion and the stellar birthline. ApJ 475(2):770–785
5. Jansen DJ, Spaans M, Hogerheijde MR, van Dishoeck EF (1995) Millimeter and submillimeter observations of the Orion Bar. II. Chemical models. A&A 303:541
6. Jansen DJ, van Dishoeck EF, Black JH, Spaans M, Sosin C (1995) Physical and chemical structure of the IC 63 nebula. II. Chemical models. A&A 302:223
7. Jørgensen JK, Visser R, Sakai N, Bergin EA, Brinch C, Harsono D, Lindberg JE, van Dishoeck EF, Yamamoto S, Bisschop SE, Persson MV (2013) A recent accretion burst in the low-mass protostar IRAS 15398–3359: ALMA imaging of its related chemistry. ApJ 779(2):L22
8. Kristensen LE, Klaassen PD, Mottram JC, Schmalzl M, Hogerheijde MR (2013) ALMA CO J = 6–5 observations of IRAS 16293–2422. Shocks and entrainment. A&A 549:L6
9. Kristensen LE, van Dishoeck EF, Bergin EA, Visser R, Yıldız UA, San Jose-Garcia I, Jørgensen JK, Herczeg GJ, Johnstone D, Wampfler SF, Benz AO, Bruderer S, Cabrit S, Caselli P, Doty SD, Harsono D, Herpin F, Hogerheijde MR, Karska A, van Kempen TA, Liseau R, Nisini B, Tafalla M, van der Tak F, Wyrowski F (2012) Water in star-forming regions with Herschel (WISH). II. Evolution of 557 GHz 1_{10}-1_{01} emission in low-mass protostars. A&A 542:A8
10. Lee C-F, Mundy LG, Reipurth B, Ostriker EC, Stone JM (2000) CO outflows from young stars: confronting the jet and wind models. ApJ 542(2):925–945
11. Lombardi M, Lada CJ, Alves J (2008) Hipparcos distance estimates of the Ophiuchus and the Lupus cloud complexes. A&A 480(3):785–792
12. Mardones D, Myers PC, Tafalla M, Wilner DJ, Bachiller R, Garay G (1997) A search for infall motions toward nearby young stellar objects. ApJ 489(2):719–733
13. Maret S, Belloche A, Maury AJ, Gueth F, André P, Cabrit S, Codella C, Bontemps S (2014) First results from the CALYPSO IRAM-PdBI survey. I. Kinematics of the inner envelope of <ASTROBJ>NGC 1333-IRAS2A</ASTROBJ>. *A&A*, 563:L1
14. Mizuno A, Fukui Y, Iwata T, Nozawa S, Takano T (June 1990) A remarkable multilobe molecular outflow: rho ophiuchi east, associated with IRAS 16293–2422. ApJ 356:184
15. Müller HSP, Schlöder F, Stutzki J, Winnewisser G (2005) The cologne database for molecular spectroscopy, CDMS: a useful tool for astronomers and spectroscopists. J Molecular Struct 742(1-3):215–227
16. Ohashi N, Hayashi M, Ho PTP, Momose M (1997) Interferometric imaging of IRAS 04368+2557 in the L1527 molecular cloud core: a dynamically infalling envelope with rotation. ApJ 475(1):211–223
17. Okoda Y, Oya Y, Sakai N, Watanabe Y, Jørgensen JK, Van Dishoeck EF, Yamamoto S (2018) The co-evolution of disks and stars in embedded stages: the case of the very-low-mass protostar IRAS 15398–3359. ApJ 864(2):L25

18. Palla F (1999) The evolution of pre-main-sequence stars. In: Lada CJ, Kylafis ND (eds) The origin of stars and planetary systems, volume 540 of NATO Advanced Study Institute (ASI) series C, p 375
19. Palla F, Stahler SW (1991) The evolution of intermediate-mass protostars. I. Basic results. ApJ 375:288
20. Sakai N, Ceccarelli C, Bottinelli S, Sakai T, Yamamoto S (2012) Distribution of CH_3OH in NGC 1333 IRAS4B. ApJ 754(1):70
21. Sakai N, Oya Y, Sakai T, Watanabe Y, Hirota T, Ceccarelli C, Kahane C, Lopez-Sepulcre A, Lefloch B, Vastel C, Bottinelli S, Caux E, Coutens A, Aikawa Y, Takakuwa S, Ohashi N, Yen H-W, Yamamoto S (2014) A chemical view of protostellar-disk formation in L1527. ApJ 791(2):L38
22. Sakai N, Sakai T, Hirota T, Burton M, Yamamoto S (2009) Discovery of the second warm carbon-chain-chemistry source, IRAS15398 - 3359 in lupus. ApJ 697(1):769–786
23. Sakai N, Sakai T, Hirota T, Watanabe Y, Ceccarelli C, Kahane C, Bottinelli S, Caux E, Demyk K, Vastel C, Coutens A, Taquet V, Ohashi N, Takakuwa S, Yen H-W, Aikawa Y, Yamamoto S (2014) Change in the chemical composition of infalling gas forming a disk around a protostar. Nature 507(7490):78–80
24. Sakai N, Sakai T, Hirota T, Yamamoto S (2008) Abundant carbon-chain molecules toward the low-mass protostar IRAS 04368+2557 in L1527. ApJ 672(1):371–381
25. Sakai N, Yamamoto S (2013) Warm carbon-chain chemistry. Chem Rev 113(12):8981–9015
26. Sternberg A, Dalgarno A (1995) Chemistry in dense photon-dominated regions. ApJS 99:565
27. Tachihara K, Dobashi K, Mizuno A, Ogawa H, Fukui Y (1996) 13CO (J= 1–0) observations of the lupus molecular clouds. PASJ 48:489–502
28. Tobin JJ, Hartmann L, Chiang HF, Wilner DJ, Looney LW, Loinard L, Calvet N, D'Alessio P (2012) A 0.2-solar-mass protostar with a Keplerian disk in the very young L1527 IRS system. Nature 492(7427):83–85. https://doi.org/10.1007/978-981-19-1708-0_5
29. van der Tak FFS, Black JH, Schöier FL, DJ, van Dishoeck EF (2007) A computer program for fast non-LTE analysis of interstellar line spectra. With diagnostic plots to interpret observed line intensity ratios. A&A 468(2):627–635
30. van Kempen TA, van Dishoeck EF, Hogerheijde MR, Güsten R (2009) Dense and warm molecular gas in the envelopes and outflows of southern low-mass protostars. A&A 508(1):259–274
31. Yıldız UA, Kristensen LE, van Dishoeck EF, Hogerheijde MR, Karska A, Belloche A, Endo A, Frieswijk W, Güsten R, van Kempen TA, Leurini S, Nagy Z, Pérez-Beaupuits JP, Risacher C, van der Marel N, van Weeren RJ, Wyrowski F (2015) APEX-CHAMP$^+$ high-J CO observations of low-mass young stellar objects. IV. Mechanical and radiative feedback. A&A 576:A109
32. Yıldız UA, Kristensen LE, van Dishoeck EF, Belloche A, van Kempen TA, Hogerheijde MR, Güsten R, van der Marel N (2012) APEX-CHAMP$^+$ high-J CO observations of low-mass young stellar objects. III. NGC 1333 IRAS 4A/4B envelope, outflow, and ultraviolet heating. A&A 542:A86

Chapter 6
IRAS 16293–2422 Source A

6.1 Introduction

In previous chapters, the analyses of the kinematic structures of the infalling-rotating envelope and the outflow have been demonstrated in the two sources: L1527 (Chap. 4) and IRAS 15398−3359 (Chap. 5). Both of them show the chemical characteristics of warm carbon-chain chemistry (WCCC), and their envelope components are traced by unsaturated carbon-chain and related molecules. In principle, the physics in the infalling-rotating envelope model itself does not depend on the chemistry of the source. Thus, another source with a chemical characteristics quite different from the above sources is addressed in this chapter: a hot corino, IRAS 16293−2422 Source A. The applicability of the infalling-rotating envelope model to this source is examined regardless of the chemical characteristics.

IRAS 16293−2422 is a well-known Class 0 protostellar binary source in Ophiuchus (d = 120 pc [17]), whose molecular gas distribution and dynamics, outflows, and chemical composition have extensively been studied (e.g. [3, 4, 6, 7, 9–11, 15, 20, 24, 26, 27, 35, 36, 38, 40, 41]). L1527 is rich in unsaturated carbon-chain molecules, whereas IRAS 16293−2422 is rich in saturated complex organic molecules (COMs), such as $HCOOCH_3$ and $(CH_3)_2O$ in the vicinity of the protostars [4, 7, 18, 30, 35]. This characteristic chemical composition is known as hot corino chemistry [4, 8]. Because of the proximity to the Sun and relatively bright emission of COMs, this source has been subject to extensive astrochemical studies. IRAS 16293−2422 consists of Source A and Source B, which are separated by 5″ [12]. The protostellar masses are reported to be $\sim$1 $M_\odot$ for Source A and < 0.1 $M_\odot$ for Source B, and it is suggested that Source B rotates around Source A [4, 6]. Since Source B has a disk/envelope component with an almost face-on configuration, its rotation motion can scarcely be detected [30, 42]. Instead, inverse P-Cygni profiles toward Source B are reported in COM lines, indicating that COMs reside in the gas

This chapter has been published in Oya et al., 2016, ApJ, 824, 88. © AAS. Reproduced with permission.

Y. Oya, *A Few Tens au Scale Physical and Chemical Structures Around Young Low-Mass Protostars*, Springer Theses, https://doi.org/10.1007/978-981-19-1708-0_6

falling to the protostar. On the other hand, Source A shows a rotation signature [30]. [14] reported a clear spin-up feature toward the protostar in the $C^{34}S$ ($J = 7 - 6$) emission observed with the SMA (Submillimeter Array) and the eSMA (Extended Submillimeter Array) at a resolution of 0.″46 × 0.″29. They suggested two possibilities as the origin of the rotation motion; a Keplerian disk and an infalling-rotating envelope. Since a CS line is found to trace the infalling-rotating envelope in L1527 (Chap. 4), it is likely that the rotation signature found by [14] can be attributed to the infalling-rotating envelope in IRAS 16293−2422 Source A. In this chapter, ALMA archival data of other molecular lines observed toward IRAS 16293−2422 Source A is analyzed to examine whether its infalling-rotating envelope and its centrifugal barrier can be identified.

6.2 Observation Data

The archival data of IRAS 16293−2422 observed in the ALMA Cycle 1 operation are used in this chapter (Project code: 2012.1.00712.S). The analyzed molecular lines are listed in Table 6.1. The 1.2 mm continuum and molecular line Images were obtained by using the CLEAN algorithm. The details of the observations are described in Appendix of this chapter. The continuum image was obtained by averaging line-free channels. After subtracting the continuum component directly from the visibilities, the line images were obtained. The beam sizes for the images are listed in Table 6.1. The rms noise levels are 2.0, 2.0, 4.0, 1.8, and 2.0 mJy beam^{-1} for the 1.2 mm continuum, OCS, CH_3OH, $HCOOCH_3$, and H_2CS images, respec-

Table 6.1 Parameters of the observed lines

Molecule	Transition	Frequency (GHz)	$E_u k_B^{-1}$ (K)	$S\mu^2$ [a] (D^2)	Beam Size
OCS[b]	$J = 19 - 18$	231.0609934	111	9.72	0″.65 × 0″.51 (P.A. 85.°29)
CH_3OH[b]	$11_{0,11} - 10_{1,10}$; A+	250.5069800	153	10.6	0″.60 × 0″.47 (P.A. 80.°34)
$HCOOCH_3$[c]	$19_{9,10} - 19_{8,11}$; E	232.5972780	166	4.10	0″.64 × 0″.51 (P.A. 85.°78)
H_2CS[b]	$7_{0,7} - 6_{0,6}$	240.2668724	46	19.0	0″.53 × 0″.46 (P.A. 73.°48)
H_2CS[b]	$7_{2,5} - 6_{2,4}$	240.5490662	99	17.5	0″.53 × 0″.46 (P.A. 73.°66)
H_2CS[b]	$7_{4,4} - 6_{4,3}$, $7_{4,3} - 6_{4,2}$	240.3321897	257	12.8, 12.8	0″.53 × 0″.46 (P.A. 73.°57)

[a]Nuclear spin degeneracy is not included
[b]Taken from CDMS [25]
[c]Taken from JPL [29]

tively, for a channel width of 122 kHz. The coordinates of the continuum peak in Source A are derived by the two-dimensional Gaussian fitting: $(\alpha_{2000}, \delta_{2000})$ = $(16^h32^m22\overset{s}{.}8713 \pm 0\overset{s}{.}0012, -24°28'36.''5023 \pm 0.''0111)$. This chapter addresses only IRAS 16293−2422 Source A to examine its kinematic structure and chemical differentiation occurring there, while Source B is addressed in Chap. 7.

6.3 Molecular Line Distribution

The ALMA data obtained toward IRAS 16293−2422 contain a number of spectral lines of various molecular species including COMs. Among them, the molecular lines with the most extended distributions are expected to be suitable to trace the infalling-rotating envelope component suggested by the $C^{34}S$ line observed with the SMA and the eSMA [14]. The OCS ($J = 19 - 18$) line is found to be the best for this purpose. Its integrated intensity map is shown in Fig. 6.1a. The size of the OCS distribution deconvolved by the beam is derived to be $(1.''665 \pm 0.''017) \times (1.''351 \pm 0.''015)$ (P.A. $21°.2 \pm 2°.2$) in FWHM (full width at half maximum) by the two-dimensional Gaussian fitting. As well, the deconvolved size of the $C^{34}S$ distribution analyzed by [14] is estimated to be $1.''6$ by using the SMA/eSMA archival data. Hence, the sizes of the OCS and $C^{34}S$ distributions are comparable to each other, and thus, the OCS emission most likely traces the infalling-rotating envelope as well as the $C^{34}S$ emission. In fact, this prediction is verified by investigating its velocity structure in the following sections (Sects. 6.4.1 and 6.6.1). It should be noted that the peak position of the $C^{34}S$ emission is offset from the dust continuum peak position (submillimeter continuum source Aa reported by [12]) by $0.''52$, while such an offset is smaller for the OCS emission ($0.''14$). In contrast, the integrated intensity maps of the CH_3OH ($11_{0,11} - 10_{1,10}$; A^+) and $HCOOCH_3$ ($19_{9,10} - 19_{8,11}$; E) lines (Fig. 6.1b, c) show more compact and centrally concentrated distributions than the OCS and $C^{34}S$ lines. Figure 6.1d–f show the integrated intensity maps of the three lines of H_2CS ($7_{0,7} - 6_{0,6}$; $7_{2,5} - 6_{2,4}$; $7_{4,4} - 6_{4,3}/7_{4,3} - 6_{4,2}$). They are also concentrated around the protostar. The sizes of the CH_3OH, $HCOOCH_3$, and H_2CS ($7_{0,7} - 6_{0,6}$) distributions deconvolved by the beam are obtained to be $(1.''042 \pm 0.''018) \times (0.''862 \pm 0.''016)$ (P.A. $46°.4 \pm 4°.5$), $(0.''728 \pm 0.''023) \times (0.''371 \pm 0.''020)$ (P.A. $58°.5 \pm 2°.5$), and $(0.''876 \pm 0.''011) \times (0.''572 \pm 0.''007)$ (P.A. $79°.3 \pm 1°.1$) in FWHM, respectively, by the two-dimensional Gaussian fitting.

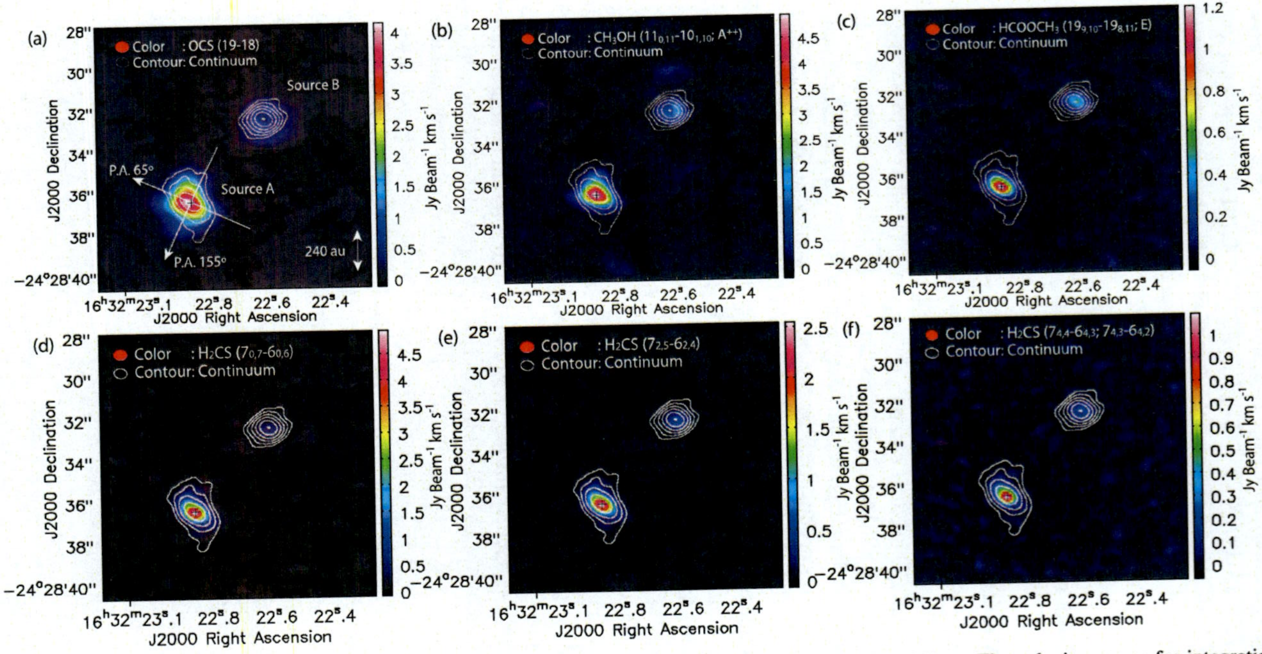

Fig. 6.1 Integrated intensity maps (moment 0 maps) of the OCS **a**, CH_3OH **b**, $HCOOCH_3$ **c**, and H_2CS **d–f** lines. The velocity ranges for integration are from −6.1 to 14.0, from −5.9 to 14.1, from −11.2 to 18.8, from −7.3 to 20.9, from −7.3 to 20.9, and from −13.2 to 11.2 km s^{-1} for panels **(a)**, **(b)**, **(c)**, **(d)**, **(e)**, and **(f)**, respectively. White contours represent the 1.2 mm continuum map. Contour levels for the continuum emission are 10, 20, 40, 80, 160, and 320σ, where the rms noise level is 2 mJy beam^{-1}. White arrows in panel **(a)** represent the envelope direction (P.A. 65°) and the outflow direction (P.A. 155°), along which the PV diagrams (Figs. 6.3, 6.5–6.13) are prepared. Black crosses represent the continuum peak position in Source A: $(\alpha_{2000}, \delta_{2000}) = (16^h32^m22^s.8713, -24°28'36.''5023)$

6.4 Velocity Structure

6.4.1 OCS

The velocity map (moment 1 map) of the OCS ($J = 19 - 18$) line is shown in Fig. 6.2a. It reveals a rotation signature around the protostar in Source A; a clear velocity gradient is found along the northeast-southwest direction. This feature is consistent with that previously reported [14, 30, 31, 36]. Here, the systemic velocity of 3.8 km s^{-1} is employed for Source A, judging from the range of the systemic velocity in the previous reports (3.6 – 3.9 km s^{-1} [4, 14, 36]). To define along which position angle (P.A.) the mid-plane of the disk/envelope system extends, the peak positions of the most red-shifted ($v_{lsr} = 10.4 - 10.7$ km s^{-1}) and blue-shifted ($v_{lsr} = -2.8 - 3.1$ km s^{-1}) OCS emission are investigated. These two components have its peak intensity at positions with slight offsets from the continuum peak position, and they and the continuum peak are well on a common line. Thus, the P.A. of the disk/envelope system is determined to be $65° \pm 10°$ (Fig. 6.1a). Assuming the outflow direction is almost perpendicular to it (i.e. $155° \pm 10°$), this result is consistent with the previous report (P.A. 144° [14]).

Figure 6.3a, b show the position-velocity (PV) diagrams of the OCS ($J = 19 - 18$) line. Their position axes are taken along the mid-plane of the disk/envelope system (hereafter 'the envelope direction') and perpendicular to it (hereafter 'the outflow direction'), which are represented in Fig. 6.1a. The PV diagram of OCS prepared along the envelope direction (Fig. 6.3a) shows a clear spin-up feature toward the protostar, while that along the outflow direction (Fig. 6.3b) shows a significant velocity gradient. The OCS emission is highly concentrated to the protostar, and thus, a contribution from outflows is unlikely. This velocity gradient is most likely the contribution from an infall motion of the rotating envelope in the vicinity of the protostar. Such a velocity gradient along the outflow direction is similar to that of the infall motion observed in the CS emission toward L1527 (Chap. 4). Therefore, the rotation motion found in the OCS emission is not Keplerian, but the gas has infall motion at a 400 au scale in diameter, as schematically illustrated in Fig. 6.4.

6.4.2 CH_3OH and $HCOOCH_3$

The velocity map (moment 1 map) of the CH_3OH ($11_{0,11} - 10_{1,10}$; A^+) line is shown in Fig. 6.2b, and it shows a clear velocity gradient around the protostar. Its PV diagram prepared along the envelope direction (Fig. 6.3c) also supports the rotation signature, as in the case of the OCS line (Sect. 6.4.1). The CH_3OH distribution is more concentrated to the protostar than the OCS distribution, as described in Sect. 6.3. On the contrary, the PV diagram of the CH_3OH line prepared along the outflow direction (Fig. 6.3d) shows little velocity gradient unlike the OCS case (Fig. 6.3b). This suggests less infall motion in the CH_3OH emission than in the OCS emission.

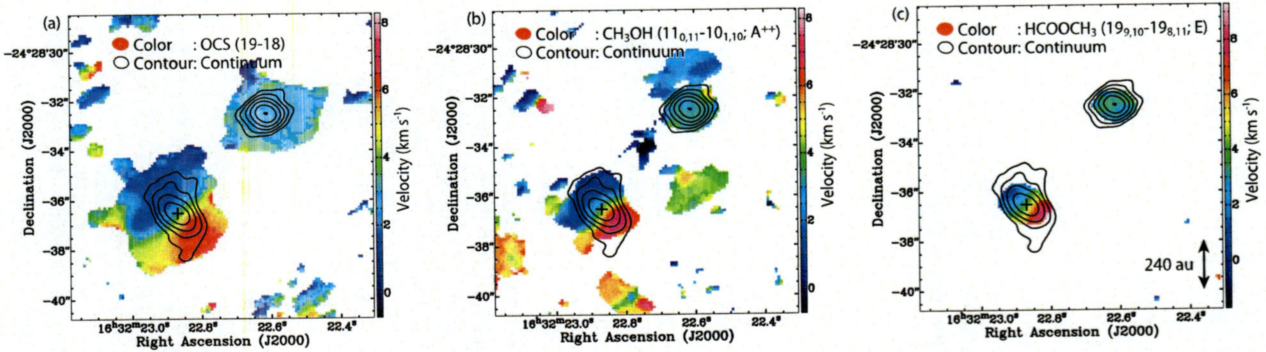

Fig. 6.2 Velocity maps (moment 1 maps) of the OCS **a**, CH_3OH **b**, and $HCOOCH_3$ **c** lines. Black contours represent the 1.2 mm continuum map, which are the same as the white contours in Fig. 6.1. Black crosses represent the continuum peak position in Source A

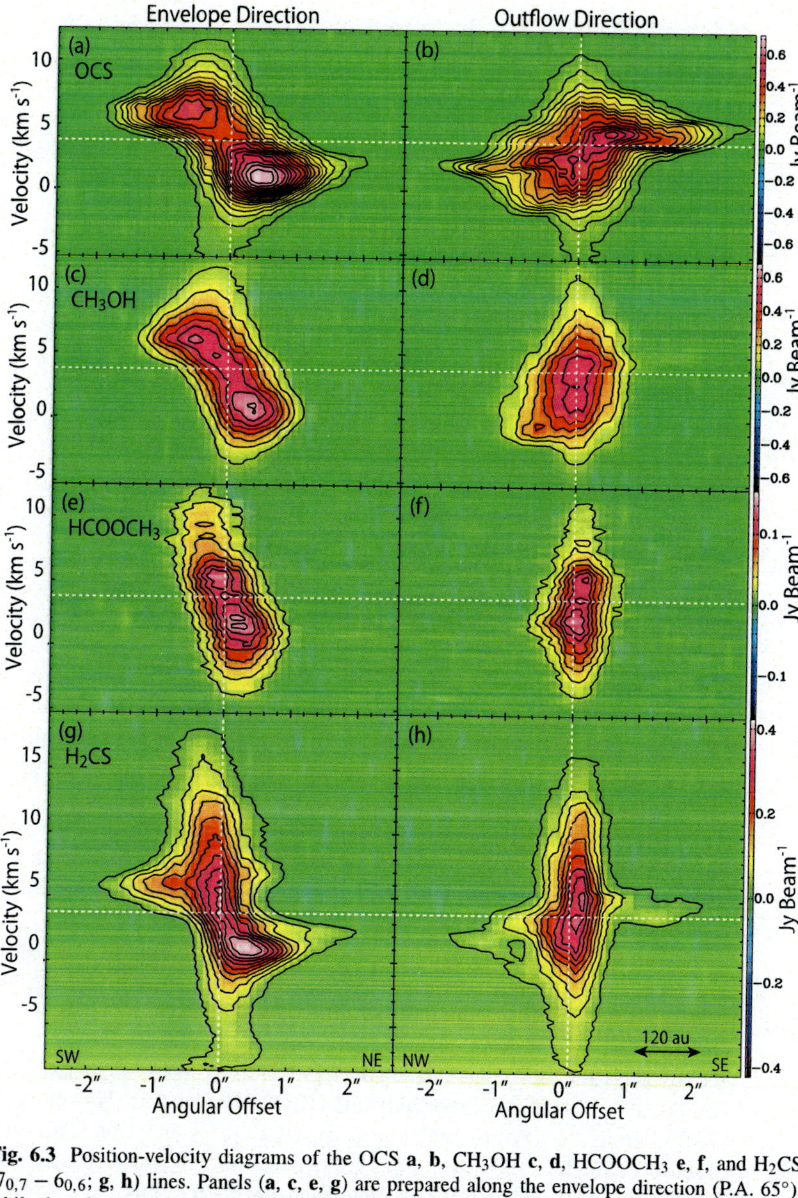

Fig. 6.3 Position-velocity diagrams of the OCS **a**, **b**, CH_3OH **c**, **d**, $HCOOCH_3$ **e**, **f**, and H_2CS ($7_{0,7} - 6_{0,6}$; **g**, **h**) lines. Panels (**a**, **c**, **e**, **g**) are prepared along the envelope direction (P.A. 65°), while the others are along the outflow direction (P.A. 155°). Their position axes are represented in Fig. 6.1a. The first contour levels are at 20σ, 20σ, 10σ, and 10σ and the level steps are 20σ, 20σ, 10σ, and 20σ for the OCS, CH_3OH, $HCOOCH_3$, and H_2CS lines, where the rms noise levels are 2.0, 4.0, 1.8, and 2.0 mJy beam^{-1}, respectively

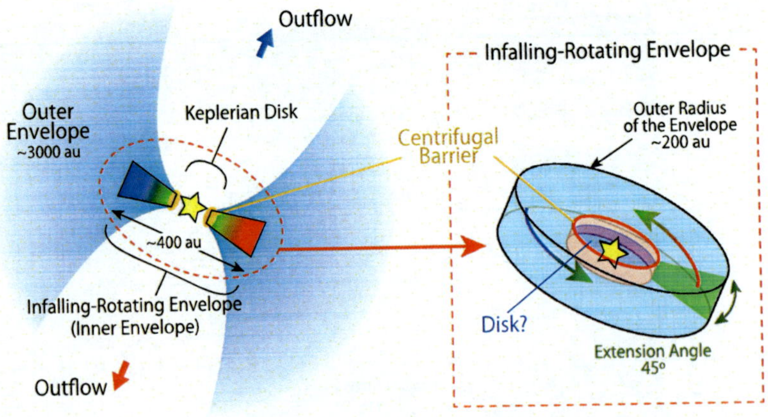

Fig. 6.4 Schematic illustration of the geometrical structure in IRAS 16293−2422 Source A. Outer envelope part in the left panel is expected to be resolved out in these ALMA observations. The kinematic structure of the inner envelope component is analyzed with the infalling-rotating envelope model in Sect. 6.5. At the most inner part inside the centrifugal barrier of the infalling-rotating envelope, the Keplerian disk component is traced by H_2CS (Fig. 6.3g, h). Its kinematic structure is investigated in Sect. 6.5.3

It is consistent with that the CH_3OH line has a more compact distribution and traces an inner part of the infalling-rotating envelope than the OCS line, because the infall velocity of the envelope gas decreases as the gas approaching to its centrifugal barrier (Chap. 3). In Fig. 6.3c (the PV diagram along the envelope direction), the intensity peaks of the CH_3OH emission with blue-shifted and red-shifted velocities appear at the positions with an angular offset of about $+0.''5$ and $-0.''5$, respectively. This feature can be explained if the CH_3OH emission mainly comes from a ring-like structure with a radius of about $0.''5$ in the innermost part of the infalling-rotating envelope.

As for the $HCOOCH_3$ ($19_{9,10} - 19_{8,11}$; E) line, its velocity map is shown in Fig. 6.2c, while its PV diagrams are in Fig. 6.3e, f. Although $HCOOCH_3$ has a more compact distribution concentrated to the protostar than CH_3OH, their overall velocity structures are essentially similar to each other. In the PV diagrams prepared along the envelope direction (Fig. 6.3c, e), the CH_3OH line shows a slight extension on the southwestern side of the protostar, while $HCOOCH_3$ does not. No velocity gradient is seen along the outflow direction in the $HCOOCH_3$ emission. This suggests an absence of infall motion in the gas from which the $HCOOCH_3$ emission mainly comes. This feature is similar to what seen in the SO emission observed toward L1527 and TMC−1A [32–34].

6.4.3 H_2CS

Figure 6.3g, h show the PV diagrams of H_2CS ($7_{0,7} - 6_{0,6}$) prepared along the envelope and outflow directions. In Fig. 6.5, they are also shown in contours superposed on those of OCS, CH_3OH, and $HCOOCH_3$. Fig. 6.1d–f shows the integrated intensity maps of the H_2CS ($7_{0,7} - 6_{0,6}$; $7_{2,5} - 6_{2,4}$; $7_{4,4} - 6_{4,3}/7_{4,3} - 6_{4,2}$) lines, and they shows compact distributions concentrated to the protostar as well as the CH_3OH and $HCOOCH_3$ cases (Fig. 6.1b, c). However, the H_2CS lines show much different features from the CH_3OH and $HCOOCH_3$ lines in their PV diagrams (Fig. 6.5). First, components with a large velocity dispersion are seen near the protostellar position in the PV diagrams of the H_2CS lines. As shown in Fig. 6.5a, b, the maximum velocity-shift from the systemic velocity (3.8 km s^{-1}) is as large as 14 km s^{-1} in the H_2CS lines, while that for the OCS line is the half ($\sim$7 km s^{-1}). This large velocity-dispersion components in H_2CS may be the contribution of the Keplerian disk component in the vicinity of the protostar. As well, the H_2CS emission shows weak emission coming from the envelope component in its PV diagram along the envelope direction. This feature is unlike the CH_3OH and $HCOOCH_3$ cases, as shown in Fig. 6.5c, e. Therefore, H_2CS likely resides in both the infalling-rotating envelope and the disk component inside its centrifugal barrier.

6.5 Infalling-Rotating Envelope Model

As described above, line emissions of different molecular species trace different physical structures in IRAS 16293−2422 Source A. In this section, their distributions and kinematic structures are analyzed by using a simple physical model (Chap. 3). Although the envelope gas of IRAS 16293−2422 extends at a 3000 au scale with a spherical configuration [9], only its innermost part at a 400 au scale is focused on in this study, which shows the rotating motion of the inner envelope and the disk around protostar (Fig. 6.4). In these observations, the outer envelope component is likely resolved out as shown in the integrated intensity maps (Fig. 6.1). Hence, the compact and flattened structure in the vicinity of the protostar is examined by comparing with cylindrical models for an envelope and a disk.

6.5.1 OCS

The velocity structure of the OCS line (Fig. 6.3a, b) suggests the existence of an infalling-rotating envelope around the dust continuum peak. Hence, it is analyzed with a simple model of an infalling-rotating envelope (Chap. 3), which has been applied to other low-mass protostellar sources (Chaps. 4, 5; TMC−1A [32]). There are two major parameters to determine the velocity field in an infalling-rotating

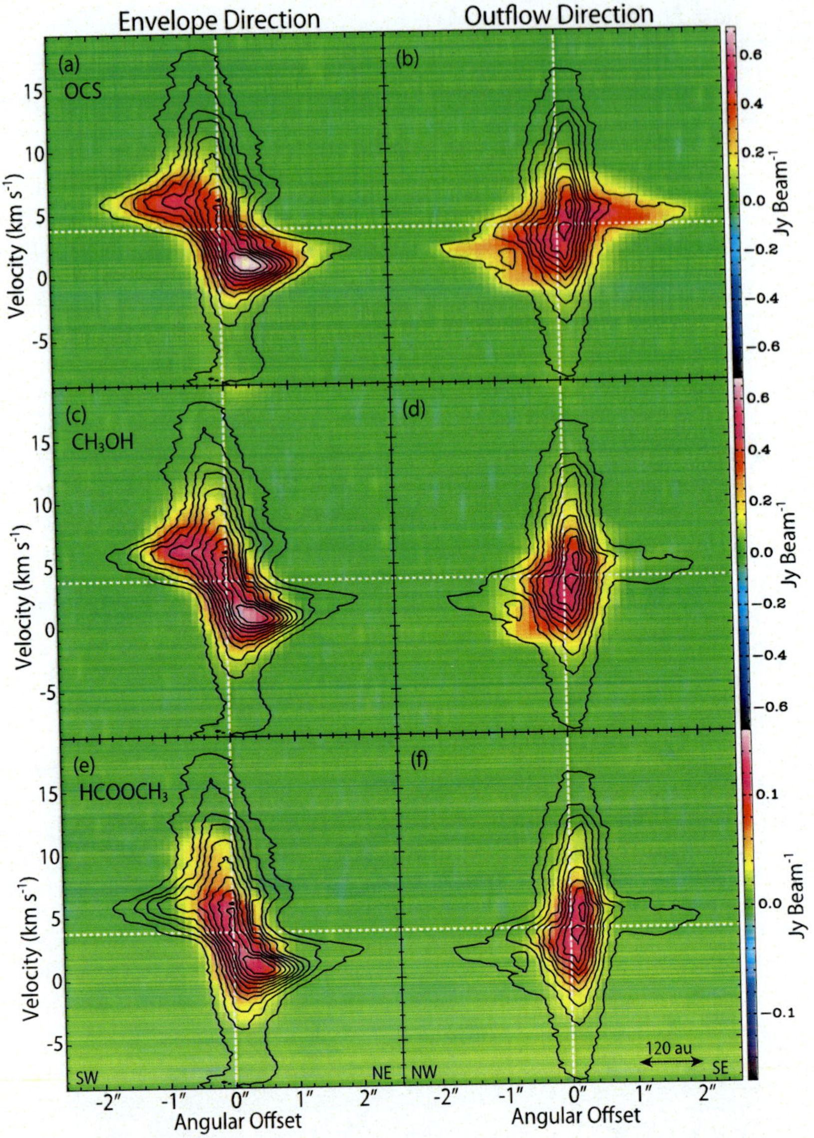

Fig. 6.5 PV diagrams of the OCS, CH_3OH, $HCOOCH_3$ (color) and H_2CS ($7_{0,7} - 6_{0,6}$; black contours) lines. Panels (**a**), (**c**), and (**e**) are prepared along the envelope direction (P.A. 65°), while others are along the outflow direction (P.A. 155°). Contour levels are every 20σ from 10σ, where the rms noise level is 2.0 mJy beam^{-1}

envelope; the protostellar mass (M) and the radius of the centrifugal barrier (r_{CB}). Besides, the infalling-rotating envelope model has additional parameters to account for the observed PV diagrams; the inclination angle of the disk/envelope system, its scale hight and/or flare angle, and its outer radius (R), as shown in Fig. 6.4 (see Chap. 3). It should be noted that R does not represent the size of the envelope gas itself but the apparent size of the distribution of an observed molecular emission. The model images are convolved by the beam size for the observed OCS line ($0.''65 \times 0.''51$; P.A. 85.°29; Table 6.1) and the intrinsic line width of 1 km s^{-1} considering possible turbulent motions in the inner envelope.

Simulations of the PV diagrams of the OCS line are conducted with a wide range of parameters. Figures 6.6 and 6.7 shows examples with various sets of M and r_{CB}; the former is for the PV diagram prepared along the envelope direction, and the latter for that along the outflow direction. These figures demonstrate how sensitive to M and r_{CB} the simulated diagrams are. Considering both the PV diagrams along two directions, the observation is reasonably reproduced by the model with M of 0.75 $M_\odot$ and r_{CB} of 50 au. Here, an inclination angle (i) of 60° (0° for a face-on configuration) and the distance of 120 pc [17] are assumed. In fact, this result is verified based on the root mean square (rms) of the residuals between the observed and the simulated diagrams (Table 6.2); the rms value for the model with the above parameters is confirmed to have almost the lowest rms value. It should be noted that there is no statistical meaning in the rms values of the residuals shown in Table 6.2. This is because a simple model is employed in this study to concentrate on the velocity structure of the envelope gas, which does not consider various quantitative effect, such as abundance variation, asymmetrical molecular distribution, optical depth effects, and so on. Therefore, the statistical noise would be overwhelmed by systematic errors caused by these assumption, which makes the chi-square analysis difficult (see also Chap. 3). Nevertheless, it should be stressed that the basic velocity structure observed in the OCS line is reasonably explained by this model, in spite of its simplicity. Further comparisons for the observed and simulated PV diagrams are performed to constrain the parameters for the model, including other parameters (e.g. an inclination angle). Then, the reasonable range for the radius of the centrifugal barrier is derived to be from 40 to 60 au. The outer radius of the distribution is obtained to be 180 au for the OCS line. The results of the best model are shown in Fig. 6.8a, b, superposed on the PV diagrams of the observed OCS line.

The protostellar mass (M) of the infalling-rotating envelope model is highly correlated with its inclination angle (i), whereas the radius of its centrifugal barrier (r_{CB}) is not. For an i ranging from 30° to 70°, an acceptable agreement is obtained between the observed and modeled PV diagrams. M which reproduces the maximum velocity-shift (4.5 km s^{-1}) due to the rotation depends on i for a given r_{CB} with the following relation:

$$\frac{M}{M_\odot} = 0.57\left(\frac{r_{CB}}{50\text{ au}}\right)\sin^{-2} i. \tag{6.1}$$

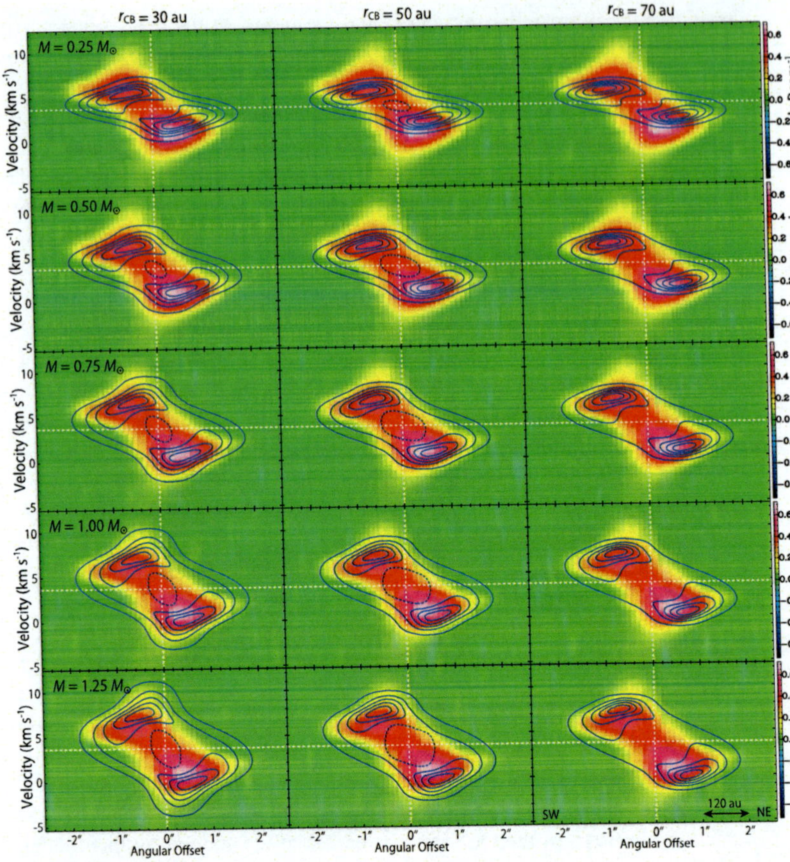

Fig. 6.6 Results of infalling-rotating envelope models (contours) superposed on the PV diagram of the OCS line (color). The position axes are taken along the envelope direction (P.A. 65°). The models are calculated for various sets of parameter values; the protostellar mass (M) and the radius of the centrifugal barrier (r_{CB}) are varied as free parameters, assuming the inclination angle (i) of 60° and the outer radius (R) of 180 au. The intrinsic line width is assumed to be 1 km s^{-1} with a Gaussian profile. Contour levels are every 20% from 5% of the peak intensity in each panel. Dashed contours around the central position in the panels for r_{CB} of 30 and 50 au stand for the intensity dip

For instance, M is obtained to be 0.97 $M_\odot$ and 0.64 $M_\odot$ for an i of 50° and 70°, respectively, with r_{CB} of 50 au. The region for M derived in this study is almost consistent with M of ~1 $M_\odot$ estimated by using the difference in the v_{lsr} between Source A and Source B [4] or the velocity width of observed lines [6]. The evaluated parameters also depends on the employed distance to IRAS 16293−2422 from the Sun (d), where M and the r_{CB} have relations with d as:

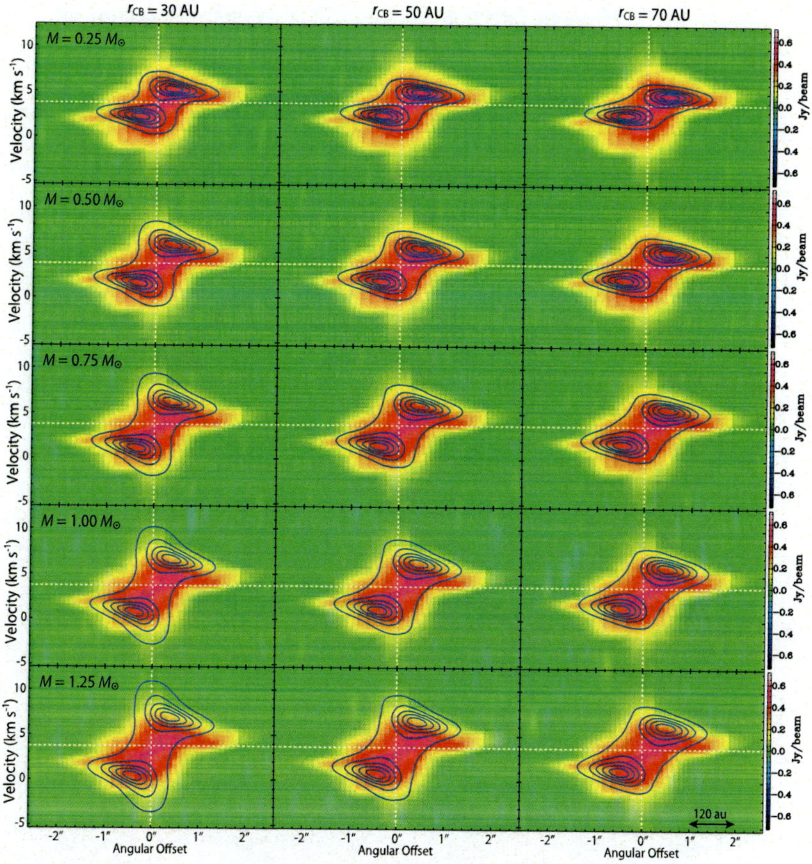

Fig. 6.7 Same as Fig. 6.6, but the position axis is taken along the outflow direction (P.A. 155°)

Table 6.2 Root mean squares of the intensity difference between the observed and calculated position-velocity diagrams of OCS[a]

	Radius of the centrifugal barrier		
Protostellar Mass	30 au	50 au	70 au
0.25 $M_\odot$	100	105	112
0.50 $M_\odot$	88	91	98
0.75 $M_\odot$	86	86	92
1.00 $M_\odot$	90	88	92
1.25 $M_\odot$	96	93	96

[a]In mJy beam^{-1}

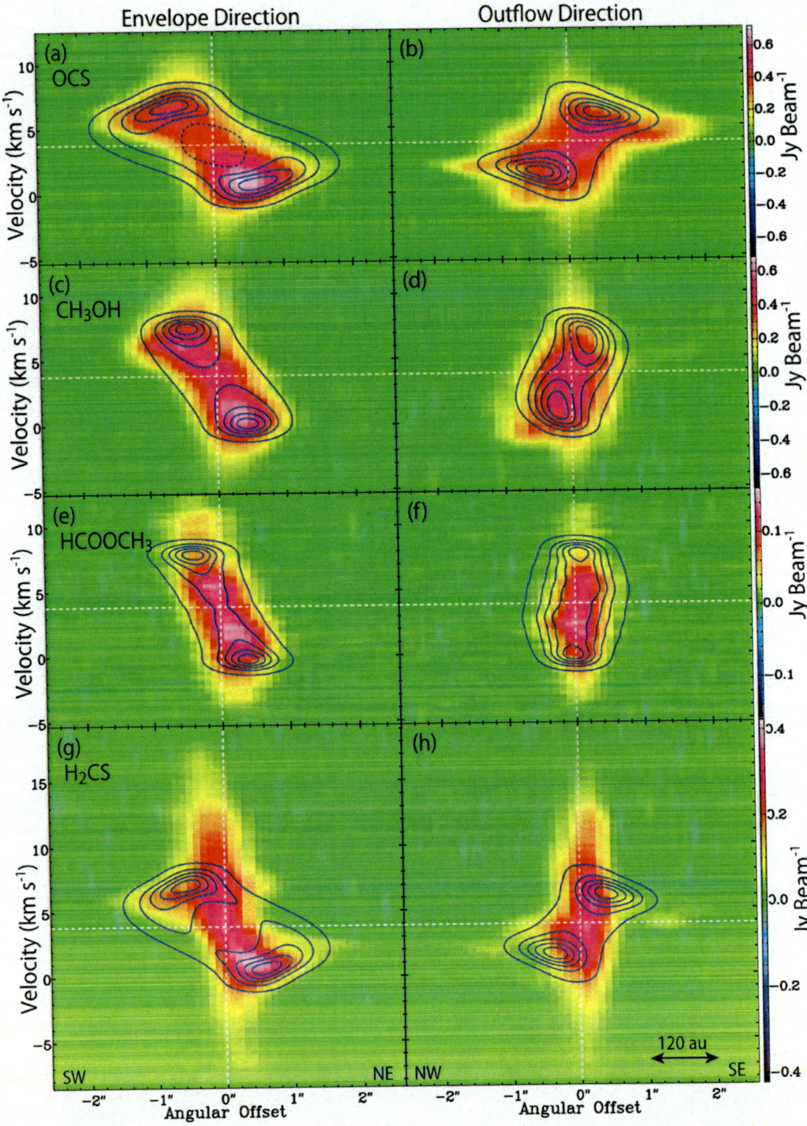

Fig. 6.8 Modeled (contours) and observed (color) PV diagrams. The observed ones are the same as those in Fig. 6.3. Blue contours represent the results of the infalling-rotating envelope models with M of 0.75 $M_{\odot}$, r_{CB} of 50 au, and i of 60°. The outer radii of the distribution (R) in the models are different among the observed molecular lines; 180, 80, 55, and 150 au for the OCS, CH_3OH, $HCOOCH_3$, and H_2CS lines, respectively. Contour levels are every 20% from 5% of the peak intensity in each panel. Dashed contour in panel (**a**) stand for the intensity dip toward the central position

$$M = 0.57\ M_\odot \times \left(\frac{d}{120\ \text{pc}}\right) \sin^{-2} i, \tag{6.2}$$

$$r_{\text{CB}} = 50\ \text{au} \times \frac{d}{120\ \text{pc}}. \tag{6.3}$$

Here, the apparent size of the centrifugal barrier ($\sim 0.''42$ in radius) and the maximum rotation velocity at the centrifugal barrier (4.5 km s^{-1}) are employed. M and r_{CB} are recalculated to be 0.65 $M_\odot$ and 57 au with d of 137 pc [28].

Figure 6.9 shows the PV diagrams of the OCS line prepared along various directions. The position axes pass the protostellar position, and they are taken for every 10° of the P.A.. Assuming the Keplerian rotation for the gas motion, the features of the PV diagrams should change symmetrically to the outflow direction (P.A. 155°); that is, the diagrams with P.A. of $(155 - n)°$ and $(155 + n)°$ shows symmetric features. However, this is not the case of the observed OCS emission. Thus, the kinematic structure traced by the OCS line cannot be explained by the Keplerian motion, but requires both rotation and infall motions. The infalling-rotating envelope model results seem to basically reproduce the observed trend for various P.A.s, although there are still some discrepancies in their detailed distributions. The discrepancies between the model and observations would originate from the asymmetric molecular distribution around the protostar seen in the integrated intensity map (Fig. 6.1a). For instance, the extension of the gas distribution is narrowest with a P.A. of 125° in the observation, while it is with 155° (outflow direction) in the model results. This asymmetrical molecular distribution cause a discrepancy between the observation and the model results larger around these P.A.s. In addition, the relative contribution of the rotation and infall motions would cause the discrepancy. By projecting the rotation and infall motions to the line of sight, they have almost opposite directions for the P.A. from 75° to 145°, and their contributions for the velocity shift cancel out each other. Thus, the PV diagrams for these P.A.s are sensitive to a small difference of their relative contribution. This situation makes difficult to reproduce the observed kinematic structure for these P.A.s.

As discussed above, the PV diagrams of the observed OCS emission cannot be explained only by the Keplerian motion. In fact, this is confirmed in Fig. 6.10, where the results of the Keplerian disk model are superposed on the observation results. Although the PV diagram prepared along the envelope direction is reproduced to some extent, the velocity gradient observed along the outflow direction cannot be explained by the Keplerian motion. Nevertheless, the possibility that OCS also resides in the Keplerian disk to some extent as well as the infalling-rotating envelope cannot be ruled out, and it would contribute to the observed features in its PV diagrams. Since this contribution is neglected in the above analysis with the infalling-rotating envelope model, it may cause additional systematic errors.

In this model, the density profile is assumed to be proportional to $r^{-1.5}$, where r denotes the distance to the protostar, and molecular abundances are assumed to be constant for simplicity. The peak intensity of the OCS line is ~ 0.7 Jy beam^{-1} for the blue-shifted velocity component. This intensity corresponds to the brightness

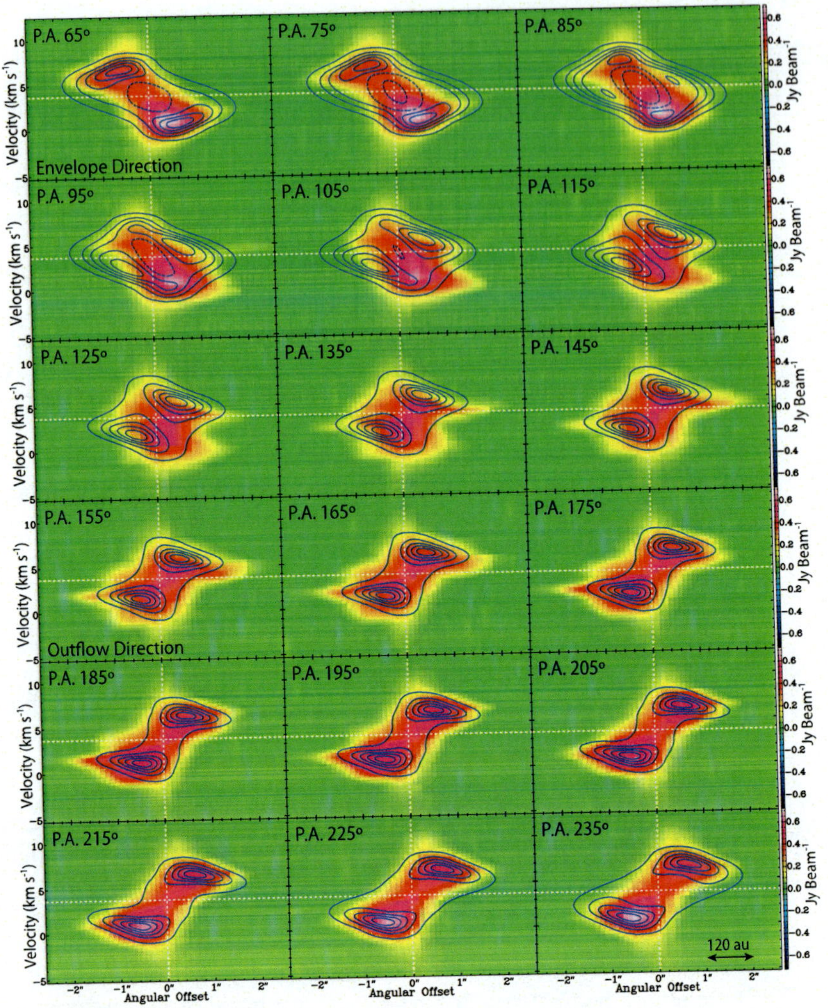

Fig. 6.9 PV diagrams of the OCS ($J = 19 - 18$; color) and the results of the infalling-rotating envelope model (blue contours). The position axes are taken along the lines passing through the protostellar position with various P.A.s, which are taken for every 10° from the envelope direction (65°). Blue contours represent the results of the infalling-rotating envelope model with M of 0.75 $M_\odot$, r_{CB} of 50 au, i of 60°, and R of 180 au, where the intrinsic line width of 1 km s^{-1} is assumed. Contour levels are every 20% from 5% of the peak intensity in each panel. Dashed contours in panels 'P.A. 65°', 'P.A. 75°', 'P.A. 85°', 'P.A. 95°', and 'P.A. 105°' stand for the intensity dip

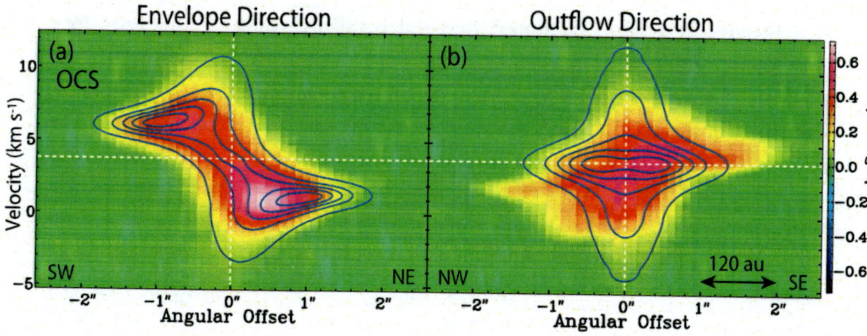

Fig. 6.10 PV diagrams of the OCS ($J = 19 - 18$; color) and the model results with the Keplerian motion (blue contours). The diagrams are prepared along the envelope direction (P.A. 65°) and the outflow direction (P.A. 155°). Blue contours represent the results of the Keplerian model with M of 1.5 $M_{\odot}$, R of 180 au, and i of 60°. Contour levels are every 20% from 5% of the peak intensity in each panel

Table 6.3 Gas kinetic temperature derived from the H_2CS lines[a]

Transitions	Red-shifted component			Blue-shifted component		
	Envelope[b]	CB[c]	Disk[d]	Envelope[b]	CB[c]	Disk[d]
$7_{0,7} - 6_{0,6}$ and $7_{2,5} - 6_{2,4}$	70 – 110	110 – 140	70 – 90	70 – 110	100 – 130	70 – 120
$7_{0,7} - 6_{0,6}$ and $7_4 - 6_4$	90 – 120	130 – 160	120 – 140	80 – 110	130 – 150	> 190

[a] In K. The gas kinetic temperatures are derived by using RADEX code [37]. The assumed ranges for the H_2 density and the column density of H_2CS are from 10^7 to 10^9 cm^{-3} and from 10^{13} to 10^{15} cm^{-2}, respectively. See footnotes of Fig. 6.14 for the error estimation

[b] Infalling-rotating envelope. The velocity-shift range for the integration is from ±1.0 to ±4.0 km s^{-1}

[c] Centrifugal barrier. The velocity-shift range for the integration is from ±2.0 to ±5.0 km s^{-1}

[d] Keplerian disk component. The velocity-shift range for the integration is from ±5.0 to ±7.3 km s^{-1}

temperature of 50 K, which is lower enough than the gas kinetic temperature about 100 K derived below (see Sect. 6.6.3 and Table 6.3). Since the excitation temperature is close to the gas kinetic temperature in the hot and dense part, this line can be assumed to be optically thin. In these observations, the OCS line is intensest, and hence, the other molecular lines are most likely to be optically thin. In order to assess how this assumption affects the results, the model calculations with density profiles of r^0 and $r^{-2.5}$ are conducted. These profiles may partly account for the effect of the optical depth, the excitation effect, and the temperature gradient effect in an effective way. However, it is found that the model results of the PV diagrams do not change significantly.

Finally, it should be noted that the infalling-rotating envelope model can also explain the essential part of the PV diagrams of the $C^{34}S$ line prepared along the envelope direction observed with the SMA and the eSMA [14], by using M and r_{CB} derived in the OCS analysis.

6.5.2 CH_3OH and $HCOOCH_3$

As well as the OCS line, the PV diagrams of the CH_3OH and $HCOOCH_3$ lines are compared with the model results (Fig. 6.8c–f). The infalling-rotating envelope models for these two molecular species have the parameter values as the same as those derived from the OCS analysis (Sect. 6.5.1), where M is 0.75 $M_\odot$, r_{CB} is 50 au, and i is 60 au, except for R. The model images are convolved by an intrinsic line width of 1 km s^{-1} and the beam size for each observed line (Table 6.1). The observed CH_3OH and $HCOOCH_3$ lines have compact distributions, as mentioned in Sect. 6.3; their sizes deconvolved by the beam are obtained to be (130 au × 100 au) and (90 au × 40 au) in FWHM, respectively, by the two-dimensional Gaussian fitting. Considering these results, R of the infalling-rotating envelope models for the CH_3OH and $HCOOCH_3$ lines are set to be 80 au and 55 au, respectively. The value for the $HCOOCH_3$ line just assumes that $HCOOCH_3$ mostly resides in a ring-like structure at the centrifugal barrier. The observed PV diagrams are reasonably reproduced by the infalling-rotating envelope models with small outer radii (Fig. 6.8c–f). More detailed comparisons are shown in Figs. 6.11 and 6.12, where the PV diagrams for every 10° of a P.A. are shown. The model results reasonably reproduces the observations, although some discrepancies are found between the observations and the model results, such as the absence of high velocity-shift components in the observed $HCOOCH_3$ line. Hence, the observed CH_3OH and $HCOOCH_3$ lines likely have emitting regions with ring-like structures around the centrifugal barrier. Such distributions of CH_3OH and $HCOOCH_3$ are similar to the SO ring around the protostar in L1527 [33, 34]. R for the CH_3OH line (80 au) is close to the centrifugal radius, which is the twice r_{CB}, possibly by chance, where the centrifugal force balances with the gravitational force.

6.5.3 H_2CS

In Fig. 6.13, the model results are overlaid on the PV diagrams of the observed three lines of H_2CS ($7_{0,7} - 6_{0,6}$; $7_{2,5} - 6_{2,4}$; $7_{4,4} - 6_{4,3}/7_{4,3} - 6_{4,2}$). Again, the parameters derived from the OCS analysis are employed for the model ($M = 0.75\ M_\odot$, $r_{CB} = 50$ au, and $i = 60°$; Sect. 6.5.1), except for R of 150 au for the H_2CS lines. The model images are convolved by an intrinsic line width of 1 km s^{-1} and the beam size for H_2CS ($7_{0,7} - 6_{0,6}$) shown in Table 6.1. Focusing on the envelope component outside the centrifugal barrier, the infalling-rotating envelope model seems to reasonably

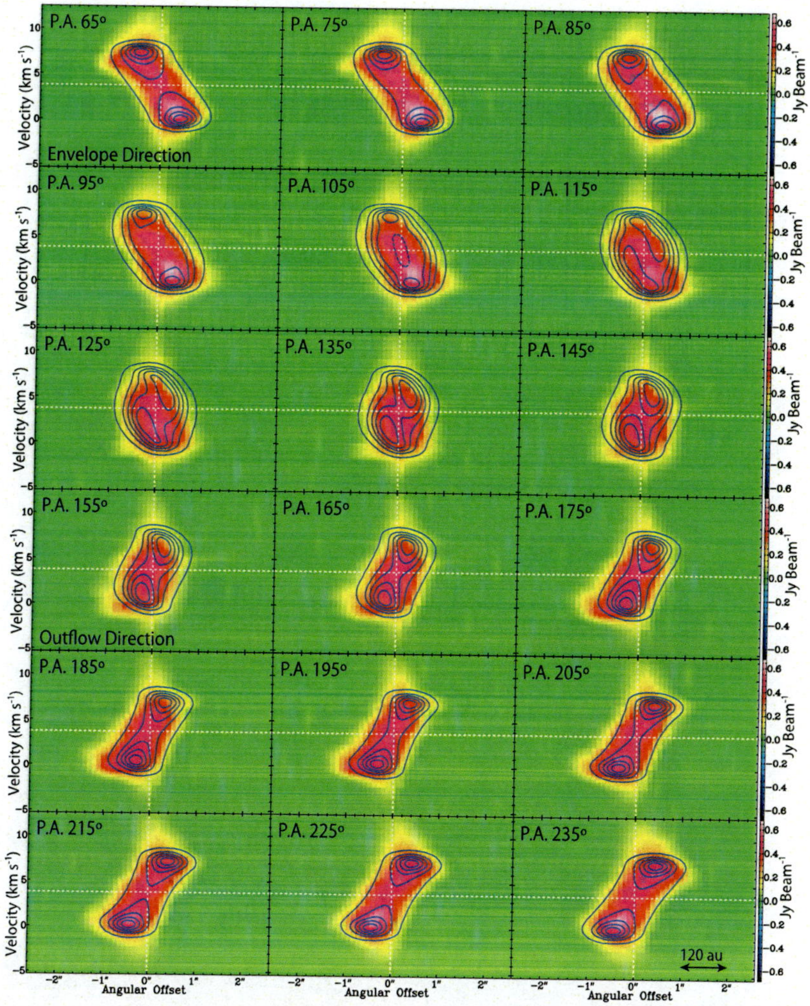

Fig. 6.11 PV diagrams of the CH_3OH ($11_{0,11} - 10_{1,10}$; A^+; color) line and the results of the infalling-rotating envelope model (blue contours). Position axes are the same as those in Fig. 6.9. Blue contours represent the results of the infalling-rotating envelope model with M of 0.75 $M_\odot$, r_{CB} of 50 au, i of 60°, and R of 80 au, where the intrinsic line width of 1 km s^{-1} is assumed. Contour levels are every 20% from 5% of the peak intensity in each panel. Dashed contours in panel 'P.A. 105°' stand for the intensity dip

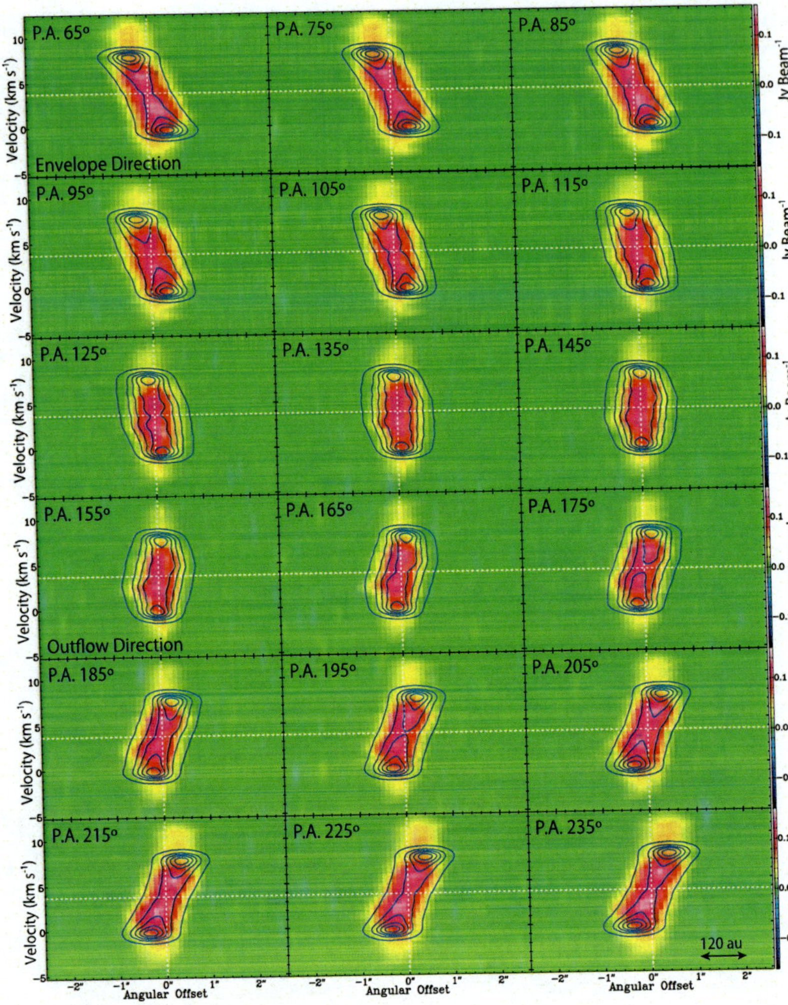

Fig. 6.12 PV diagrams of the $HCOOCH_3$ ($19_{9,10} - 19_{8,11}$; E; color) line and the results of the infalling-rotating envelope model (blue contours). Position axes are the same as those in Fig. 6.9. Blue contours represent the results of the infalling-rotating envelope model with M of 0.75 $M_{\odot}$, $r_{\rm CB}$ of 50 au, i of 60°, and R of 55 au, where the intrinsic line width of 1 km s^{-1} is assumed. Contour levels are every 20% from 5% of the peak intensity in each panel

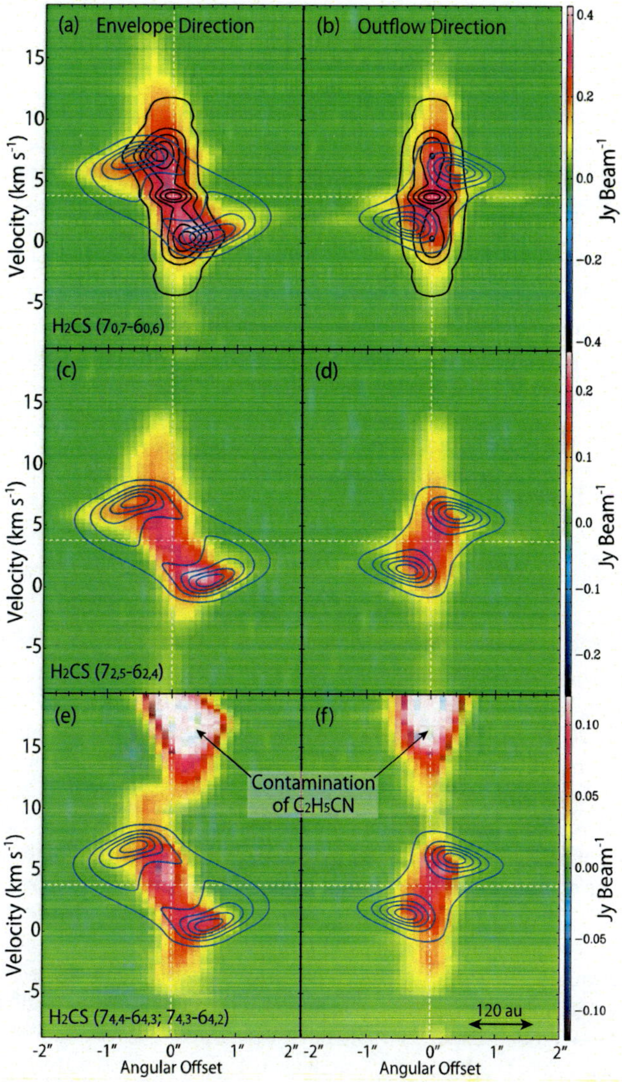

Fig. 6.13 PV diagrams of the three lines of H_2CS ($7_{0,7} - 6_{0,6}$; $7_{2,5} - 6_{2,4}$; $7_{4,4} - 6_{4,3}/7_{4,3} - 6_{4,2}$) and the model results (contours). Position axes are taken along the envelope direction (P.A. 65°) and the outflow direction (P.A. 155°). Blue contours represent the results of the infalling-rotating envelope model with M of 0.75 $M_\odot$, r_{CB} of 50 au, i of 60°, and R of 150 au, where the intrinsic line width of 1 km s^{-1} is assumed. Black contours in panels (**a**) and (**b**) represent the model results with the Keplerian motion within the centrifugal barrier, where M is 0.75 $M_\odot$ and i is 60°. Contour levels for each model are every 20% from 5% of its peak intensity in each panel. The H_2CS ($7_{4,4} - 6_{4,3}$ and $7_{4,3} - 6_{4,2}$) line is contaminated by the C_2H_5CN ($28_{1,28} - 27_{1,27}$; 240.3193373 GHz) line in panels (**e**) and (**f**)

reproduce the observation, although the distribution of the H_2CS ($7_{4,4} - 6_{4,3}/7_{4,3} - 6_{4,2}$) line tends to be less extended than the other two H_2CS lines as like as the CH_3OH and $HCOOCH_3$ cases.

On the other hand, the observed emission apparently shows a large velocity-dispersion near the protostar, which is not explained by the infalling-rotating envelope model. These broad-velocity components are likely attributed to the Keplerian disk component in the vicinity of the protostar. Therefore, the model calculations with the Keplerian motion are performed. The results are shown in Fig. 6.13a, b overlaid on the observed H_2CS ($7_{0,7} - 6_{0,6}$) emission, where M and i derived from the OCS analysis are employed (0.75 $M_{\odot}$, 60°; Sect. 6.5.1). Here, R of the Keplerian disk is assumed to be equal to the radius of the centrifugal barrier (50 au). The model results well explain the broad-velocity components, which strongly supports the existence of the Keplerian disk structure within the centrifugal barrier.

6.6 Discussion

6.6.1 *Infalling-Rotating Envelope and Its Centrifugal Barrier*

The above results most likely suggest the existence of the infalling-rotating envelope and indicate its centrifugal barrier in the prototypical hot corino source IRAS 16293–2422 Source A. The kinematic structures traced by the OCS, CH_3OH, $HCOOCH_3$, and H_2CS lines are reasonably reproduced by the simple ballistic model of an infalling-rotating envelope (Figs. 6.8, 6.9, 6.11–6.13), although the existence of its centrifugal barrier is not so clear as the L1527 case (Chap. 3; [34]) due to the contamination by the Keplerian disk component. Nevertheless, its radius can be estimated by using the ballistic model. For the identification of the centrifugal barrier in this source, the OCS emission is found to be useful. Although L1527 and IRAS 16293–2422 are quite different from each other in the chemical composition, the physical/kinematic structures in these sources are found to be similar. Therefore, an infalling-rotating envelope and its centrifugal barrier would be a common occurrence in young low-mass protostellar sources regardless of their chemical characteristics.

On the other hand, the radius of a centrifugal barrier is different among sources (Chaps. 3–5; [32]). Its empirical range is from a few tens au to 100 au in radius. This variation is discussed in Chapter 10. Therefore, statistical studies for the radius of a centrifugal barrier for a number of young protostellar sources are interesting. It will provide us with rich information regarding the formation process of protoplanetary disks and their diversity.

6.6.2 *Origin of the Chemical Change Around the Centrifugal Barrier*

In the above analysis, the chemical differentiation is found in the closest vicinity of the protostar in the hot corino source IRAS 16293−2422 Source A. A chemical change is likely occurring around the centrifugal barrier, probably between it and the centrifugal radius. This feature is similar to the case of the WCCC sources L1527 and TMC−1A (Chap. 4; [32]). In IRAS 16293−2422 Source A, the OCS emission mainly traces the infalling-rotating envelope up to the radius of 180 au, while the CH_3OH and $HCOOCH_3$ emissions are concentrated to the centrifugal barrier. Meanwhile, the H_2CS emissions trace both the infalling-rotating envelope and the Keplerian disk component within it.

The OCS emission has been detected toward hot cores and hot corinos so far [2, 3, 6]. IRAS 16293−2422 is known to be rich in OCS, because even the high-excitation lines of its isotopologues ($OC^{34}S$ and $O^{13}CS$) are detected by single-dish observations [3, 6]. According to the jump-model analysis by [35], OCS has the fractional abundance as high as 10^{-7} relative to H_2. However, its production processes are not well understood. The gas phase production (CS + OH, SO + CH) as well as the solid phase production are proposed [19, 39]. From the surface binding energy of 2888 K (UMIST database for astrochemistry [22]; http://udfa.ajmarkwick.net/index.php), the evaporation temperature of OCS is evaluated to be ~50 K (see Appendix A). This temperature is lower than the gas temperature of the infalling-rotating envelope derived from the multiple line analysis of H_2CS below, and thus, the contribution of the liberation of OCS from dust grains cannot be ruled out for its distribution in the infalling-rotating envelope.

On the other hand, the CH_3OH and $HCOOCH_3$ emission have the compact distribution concentrated to the centrifugal barrier possibly with ring-like structures. It suggests that they would be liberated from ice mantles to the gas phase due to weak accretion shocks expected around the centrifugal barrier [23]. This situation is similar to the SO distribution in the WCCC sources L1527 and TMC−1A [32–34]. Alternatively, CH_3OH and $HCOOCH_3$ would be thermally evaporated by protostellar heating in the vicinity of the protostar. Their evaporation radii could be just outside the centrifugal barrier by chance; in this case, their observed distributions can be explained. The luminosity of IRAS 16293−2422 is as high as 22 $L_\odot$ [13], and thus the evaporation region of CH_3OH would extend outward of the centrifugal barrier. The gas kinetic temperature at the radius of 50 au is estimated to be as high as 130 K according to the spherical model reported by [13], and R of the infalling-rotating envelope model for CH_3OH (80 au) coincides with the radius of the water sublimation ($\gtrsim$ 100 K). Therefore, thermal evaporation of CH_3OH and $HCOOCH_3$ may cause enhancements in their emission.

6.6.3 Gas Kinetic Temperatures of H_2CS

In order to assess the above two possibilities, the gas kinetic temperatures are derived by using the three lines of H_2CS ($7_{0,7} - 6_{0,6}$; $7_{2,5} - 6_{2,4}$; $7_{4,4} - 6_{4,3}/7_{4,3} - 6_{4,2}$). These transitions can be used as a good thermometer, because the cross K-ladder radiative transitions (i.e. $K_a = 2 \rightarrow 0$) are very slow. The results are summarized in Table 6.3, which are evaluated by the method described below.

Since H_2CS is distributed from the infalling-rotating envelope to the Keplerian disk component, the gas kinetic temperatures are derived in the infalling-rotating envelope component, at the centrifugal barrier, and in the Keplerian disks based on the integrated intensity ratios between the three H_2CS lines. The integrated intensities are obtained in circular areas with a diameter of 0.″5, which are centered at the positions with certain offsets from the continuum peak position along the envelope direction (Fig. 6.1a). Offsets of 0.″0, 0.″5, and 1.″5 are employed for the envelope component, the centrifugal barrier, and the Keplerian disk component, respectively. The diameter of 0.″5 is taken to be comparable to the bean sizes in the observation. The velocity-shift range for the integration is from ± 1.0 to ± 4.0 km s^{-1}, from ± 2.0 to ± 5.0 km s^{-1}, and from ± 5.0 to ± 7.3 km s^{-1}, for the envelope component, the centrifugal barrier, and the Keplerian disk component, respectively. It should be noted that the systemic velocity component is excluded from this analysis to avoid the self-absorption effect. For the Keplerian disk component, only the emission with a velocity-shift higher than the maximum rotation velocity in the infalling-rotating envelope is used to exclude the contamination by the envelope component.

The integrated intensity ratios between the observed H_2CS lines are compared with those calculated by using RADEX code [37] (Fig. 6.14). In the calculation with the RADEX code, the H_2 density is assumed to be from 10^7 to 10^9 cm^{-3}. The column density of H_2CS is assumed to be from 10^{13} to 10^{15} cm^{-2}, which is consistent with the previous report (3.7×10^{13} cm^{-2} [3]) considering the beam filling factor in their observation. In Fig. 6.14, the gas kinetic temperatures are calculated for the various intensity ratios of the H_2CS lines as a function of the H_2 densities (Fig. 6.14a) and the column densities of H_2CS (Fig. 6.14b). Within the above ranges for the H_2 density and the column density of H_2CS, the gas kinetic temperature barely depends on these parameters. Hence, the gas kinetic temperature can be well evaluated from the observed intensity ratios regardless of the other parameters. The gas kinetic temperatures thus evaluated are summarized in Table 6.3.

The derived gas kinetic temperatures are almost consistent with the temperature at a 50 au scale in the spherical model by [13]. Meanwhile, they are higher than the rotation temperatures of H_2CS (60 K) and SO_2 (95 K) derived from the single-dish data [3]. Moreover, the gas kinetic temperatures derived from the H_2CS ($7_{0,7} - 6_{0,6}$; $7_{2,5} - 6_{2,4}$) lines seem to increase at the centrifugal barrier. As well, that derived from the H_2CS ($7_{0,7} - 6_{0,6}$; $7_{4,4} - 6_{4,3}/7_{4,3} - 6_{4,2}$) lines shows a similar trend of the enhanced temperature at the centrifugal barrier for the red-shifted components, although that derived for the blue-shifted components has a higher temperature in the Keplerian disk component than at the centrifugal barrier. These results suggest

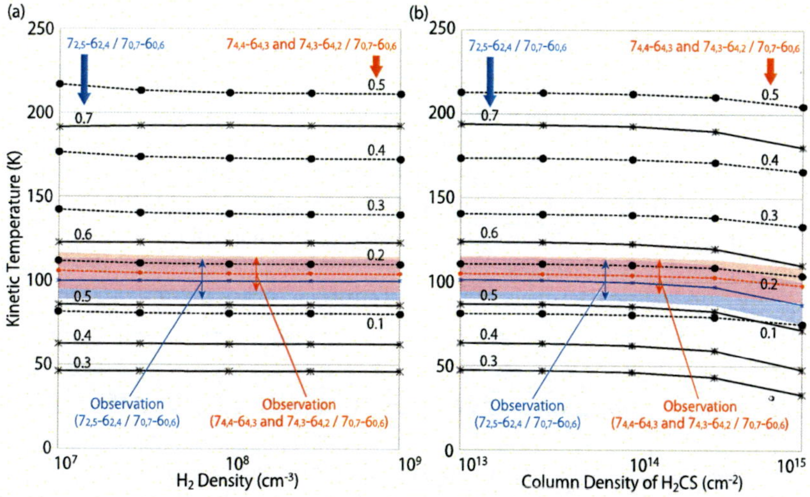

Fig. 6.14 An example of evaluation of the gas kinetic temperature by using the integrated intensity ratios of the H_2CS ($7_{2,5} - 6_{2,4}$; $7_{4,4} - 6_{4,3}$ and $7_{4,3} - 6_{4,2}$) lines relative to the H_2CS ($7_{0,7} - 6_{0,6}$) line. Solid and dashed lines represent the gas kinetic temperature for a given ratio of $7_{2,5} - 6_{2,4}/7_{0,7} - 6_{0,6}$ and ($7_{4,4} - 6_{4,3}$ and $7_{4,3} - 6_{4,2}$)/$7_{0,7} - 6_{0,6}$, respectively, as a function of the H_2 density **a** and the H_2CS column density **b** calculated by using RADEX code [37]. As demonstrations, the column density of H_2CS is fixed to be 10^{14} cm^{-2} in panel (**a**), while the H_2 density is to be 10^8 cm^{-3} in panel (**b**). Solid blue and dashed red lines represent the observed results of the $7_{2,5} - 6_{2,4}/7_{0,7} - 6_{0,6}$ and ($7_{4,4} - 6_{4,3}$ and $7_{4,3} - 6_{4,2}$)/$7_{0,7} - 6_{0,6}$ ratio, respectively, whose error ranges (3σ) are shown by the blue-colored and red-colored areas. Only the results for the red-shifted components, in the infalling-rotating envelope is shown here as examples, while the results for the other components are summarized in Table 6.3. Error ranges are estimated only from the statistical error and do not contain the calibration error, which is expected to be almost canceled out in the integrated intensity ratios

a higher gas kinetic temperature at the centrifugal barrier than in the other components. If the dust temperature is enhanced at the centrifugal barrier as well as the gas kinetic temperature, the enhancement of CH_3OH and $HCOOCH_3$ abundances around the centrifugal barrier would be expected. Moreover, these species will be depleted onto dust grains in the mid-plane of the Keplerian disk, if the dust/gas temperatures are as low as (70 – 90) K there. This mechanism may account for why the CH_3OH and $HCOOCH_3$ emissions observed in the Keplerian disk are not as intense as at the centrifugal barrier, i.e. they come from ring-like structures around the centrifugal barrier. On the other hand, the deficiency of OCS and $C^{34}S$ in the Keplerian disk component is puzzling in this context, because their binding energies (2888 K for OCS; 1900 K for CS) are comparable to that of H_2CS (2700 K) (UMIST database for astrochemistry [22]; http://udfa.ajmarkwick.net/index.php). Therefore, gas-phase destruction mechanisms of OCS and CS should be considered carefully in the disk component.

The temperature analysis described above supports an accretion shock for liberation of organic molecules at the centrifugal barrier. Alternatively, the protostellar heating combined with the geometrical effect may raise the gas/dust temperature. The envelope gas would be extended perpendicular to its mid-plane if the infall gas is stagnated near the centrifugal barrier, resulting in a large scale height there. If this is the case, the protostar can directly heats this component without shielding by the Keplerian disk, and thus, the gas kinetic temperature can be higher near the centrifugal barrier than other components. Discrimination of the above two possibilities is left for future studies, since the currently available data has insufficient spatial resolutions to investigate the vertical structure of the envelope.

6.6.4 Abundance of $HCOOCH_3$ Relative to CH_3OH

Judging from Fig. 6.8c–f, CH_3OH and $HCOOCH_3$ mainly reside in the envelope component and/or around the centrifugal barrier. As well, they would partly exist in the Keplerian disk. Thus, to evaluate their column densities, it is required to assess the contribution from each component securely. Therefore, their column densities are derived from their integrated intensities with limited velocity ranges, as the H_2CS analysis (Sect. 6.6.3). It should be noted that the derived column densities are not the total ones, and thus do not have quantitative meanings by themselves. Nevertheless, the abundance ratios between CH_3OH and $HCOOCH_3$ are still meaningful for mutual comparison. Table 6.4 shows the derived $HCOOCH_3/CH_3OH$ abundance ratio for each component. In the evaluation of the column densities, the LTE (local thermodynamic equilibrium) approximation is employed. The rotation temperatures of 100 K, 130 K, and 100 K is assumed for the envelope, the centrifugal barrier, and the Keplerian disk component, respectively, based on the gas kinetic temperatures derived from the H_2CS ($7_{0,7} - 6_{0,6}$; $7_{2,5} - 6_{2,4}$) lines (Table 6.3).

If the CH_3OH emission is assumed to be optically thin, the $HCOOCH_3/CH_3OH$ abundance ratio is found to be as high as unity or even higher, except for the red-

Table 6.4 $HCOOCH_3/CH_3OH$ Column Density Ratios[a]

	Red-shifted component			Blue-shifted component		
	Envelope[b]	CB[c]	Disk[d]	Envelope[b]	CB[c]	Disk[d]
Ratio	0.8 ± 0.5	4.2 ± 0.3	8.7 ± 1.3	2.7 ± 1.2	4.9 ± 0.3	8.9 ± 1.2

[a]Quoted errors represent 3σ, where σ is derived from the statistical error

[b]Infalling-rotating envelope. The velocity-shift range for the integration is from ±1.0 to ±4.0 km s^{-1}. The rotational temperature is assumed to be 100 K (Table 6.3)

[c]Centrifugal barrier. The velocity-shift range for the integration is from ±2.0 to ±5.0 km s^{-1}. The rotational temperature is assumed to be 130 K (Table 6.3)

[d]Keplerian disk component. The velocity-shift range for the integration is from ±5.0 to ±7.3 km s^{-1}. The rotational temperature is assumed to be 100 K (Table 6.3)

shifted component of the envelope. Such a high ratio is consistent with the previous report [5]. More importantly, the ratio likely increases from the envelope to the Keplerian disk component. This trend is commonly seen in both the red-shifted and blue-shifted components. It cannot simply be attributed to the evaporation process, because CH_3OH (4930 K) and $HCOOCH_3$ (4000 K) have almost comparable surface binding energies (UMIST database for astrochemistry [22]; http://udfa.ajmarkwick.net/index.php). Thus, this result may suggest that the chemical composition of grain mantles is processed to enhance $HCOOCH_3$ as grains pass across the centrifugal barrier. Alternatively, $HCOOCH_3$ may be formed in the gas phase around the centrifugal barrier and inside it [1].

6.7 Summary of This Chapter

In this chapter, the OCS, CH_3OH, $HCOOCH_3$, and H_2CS emissions observed toward IRAS 16293−2422 Source A are analyzed. The observations were carried out with ALMA during its Cycle 1 period at a sub-arcsecond resolution ($\sim 0.''6 \times 0.''5$). Major findings are summarized below.

(1) The molecular distributions are different among molecular species (Fig. 6.3). OCS is detected in the envelope, while CH_3OH and $HCOOCH_3$ are enhanced near the centrifugal barrier. Meanwhile, H_2CS resides in both the envelope and the Keplerian disk component.
(2) The infalling-rotating envelope model successfully reproduce the kinematic structure traced by the OCS emission (Fig. 6.9). Reasonable parameters for the model are the protostellar mass (M) of 0.75 $M_\odot$ and the radius of the centrifugal barrier of 50 au, assuming the inclination angle (i) of 60° (0° for a face-on configuration) and the distance of 120 pc to the Sun [17].
(3) The compact distributions of CH_3OH and $HCOOCH_3$ concentrated around the centrifugal barrier can be attributed to their liberation by weak accretion shocks in front of the centrifugal barrier and/or by protostellar heating (Figs. 6.11, 6.12).
(4) As well as the infalling-rotating envelope component, the H_2CS emission shows high velocity-shift components concentrated toward the protostar position (Fig. 6.5), which seem to trace the Keplerian disk component. In fact, the kinematic structures of these components can be explained by using the disk model with the Keplerian motion; the parameters for the model are the same as those for the OCS analysis (M of 0.75 $M_\odot$; i of 60°; Fig. 6.13). Based on the integrated intensities of the three lines of H_2CS ($7_{0,7} - 6_{0,6}$; $7_{2,5} - 6_{2,4}$; $7_{4,4} - 6_{4,3}/7_{4,3} - 6_{4,2}$), the gas kinetic temperatures are evaluated for the infalling-rotating envelope, the centrifugal barrier, and the Keplerian disk component (Table 6.3). There is a hint that the gas kinetic temperature is locally enhanced near the centrifugal barrier.

(5) The $HCOOCH_3/CH_3OH$ abundance ratios are evaluated for the infalling-rotating envelope, the centrifugal barrier, and the Keplerian disk component (Table 6.4). They are found to be as high as unity or even higher. More interestingly, the ratio tends to increase from the envelope to the Keplerian disk.
(6) A drastic chemical change is thus found to be occurring across the centrifugal barrier.

The above results indicate that the centrifugal barrier plays a crucial role in hot corino chemistry. Considering that the existence of centrifugal barriers has been reported for the WCCC sources (Chapter 4; [32–34]), it seems to be a common occurrence in low-mass protostellar sources regardless of their chemical characteristics.

Appendix: ALMA Data

The ALMA observations toward IRAS 16293–2422 were carried out during its Cycle 1 operations [16] on 22 May 2014, 14 June 2014, and 25 April 2014 for the frequency ranges from 230 to 250 GHz (higher frequency setting), from 220 to 240 GHz (lower frequency setting), and both of them, respectively. Spectral lines of OCS, CH_3OH, $HCOOCH_3$, and H_2CS were observed with the Band 6 receiver at a frequency of 230, 250, 230, and 240 GHz, respectively. The number of antennas used in the observations are from 42 to 44. The baseline length ranges from 17.14 to 636.53 m and from 19.58 to 628.65 m for the observations with the higher and lower frequency settings, respectively. The field center of the observations was $(\alpha_{2000}, \delta_{2000}) = (16^h32^m22^s.72, -24°28'34.''3)$. The primary beams (half-power beam width) are 24.″26 and 25.″06 for the observations with the higher and lower frequency settings, respectively. The total on-source time were 51 minutes and 50 minutes for the higher and lower frequency settings, with typical system temperatures of 200 − 300 K and 50 − 100 K, respectively. The backend correlator was tuned to a resolution of 122 kHz, corresponding to the velocity resolution of 0.15 km s^{-1} at 240 GHz, and a bandwidth of 468.750 MHz. The phase calibration was performed with J1626−2951, J1700−2610 or J1625−2527 for every 8 minutes, and the bandpass calibration was with the quasars J1733−1304, J1700−2610, and J1517−2422. The absolute flux density scale was derived from Titan. The data calibration was performed in the antenna-based manner, and the uncertainties in the observed intensities are expected to be less than 10% (ALMA Cycle 1 Technical Handbook [21]). The Brigg's weighting with the robustness parameter of 0.5 was employed to obtain the images of the continuum and the spectral lines. In this analysis, the self-calibration was not applied.

References

1. Balucani N, Ceccarelli C, Taquet V (2015) Formation of complex organic molecules in cold objects: the role of gas-phase reactions. MNRAS 449:L16–L20
2. Blake GA, Sutton EC, Masson CR, Phillips TG (1987) Molecular abundances in OMC-1: The chemical composition of interstellar molecular clouds and the influence of massive star formation. ApJ 315:621
3. Blake GA, van Dishoeck EF, Jansen DJ, Groesbeck TD, Mundy LG (1994) Molecular abundances and low-mass star formation. I. Si- and S-bearing species toward IRAS 16293-2422. ApJ 428:680
4. Bottinelli S, Ceccarelli C, Neri R, Williams JP, Caux E, Cazaux S, Lefloch B, Maret S, Tielens AGGM (2004) Near-arcsecond resolution observations of the hot Corino of the solar-type protostar IRAS 16293–2422. ApJ 617(1):L69–L72
5. Bottinelli S, Ceccarelli C, Williams JP, Lefloch B (2007) Hot corinos in NGC 1333-IRAS4B and IRAS2A. A&A 463(2):601–610
6. Caux E, Kahane C, Castets A, Coutens A, Ceccarelli C, Bacmann A, Bisschop S, Bottinelli S, Comito C, Helmich FP, Lefloch B, Parise B, Schilke P, Tielens AGGM, van Dishoeck E, Vastel C, Wakelam V, Walters A (2011) TIMASSS: the IRAS 16293-2422 millimeter and submillimeter spectral survey. I. Observations, calibration, and analysis of the line kinematics. A&A 532:A23
7. Cazaux S, Tielens AGGM, Ceccarelli C, Castets A, Wakelam V, Caux E, Parise B, Teyssier D (2003) The hot core around the low-mass protostar IRAS 16293–2422: Scoundrels Rule! ApJ 593(1):L51–L55
8. Ceccarelli C (2004) The hot corinos of solar type protostars. In: Johnstone D, Adams FC, Lin DNC, Neufeeld DA, Ostriker EC (eds) Star formation in the interstellar medium: in honor of David Hollenbach, volume 323 of astronomical society of the pacific conference series, p. 195
9. Ceccarelli C, Castets A, Caux E, Hollenbach D, Loinard L, Molinari S, Tielens AGGM (2000) The structure of the collapsing envelope around the low-mass protostar IRAS 16293–2422. A&A 355:1129–1137
10. Ceccarelli C, Loinard L, Castets A, Faure A, Lefloch B (2000) Search for glycine in the solar type protostar IRAS 16293–2422. A&A 362:1122–1126
11. Ceccarelli C, Loinard L, Castets A, Tielens AGGM, Caux E (2000) The hot core of the solar-type protostar IRAS 16293–2422: H_2CO emission. A&A 357:L9–L12
12. Chandler CJ, Brogan CL, Shirley YL, Loinard L (2005) IRAS 16293–2422: proper motions, jet precession, the hot core, and the unambiguous detection of infall. ApJ 632(1):371–396
13. Crimier N, Ceccarelli C, Maret S, Bottinelli S, Caux E, Kahane C, Lis DC, Olofsson J (2010) The solar type protostar IRAS16293-2422: new constraints on the physical structure. A&A 519:A65
14. Favre C, Jørgensen JK, Field D, Brinch C, Bisschop SE, Bourke TL, Hogerheijde MR, Frieswijk WWF (2014) Dynamical structure of the inner 100 AU of the deeply embedded protostar IRAS 16293-2422. ApJ 790(1):55
15. Jørgensen JK, Bourke TL, Nguyen Luong Q, Takakuwa S (2011) Arcsecond resolution images of the chemical structure of the low-mass protostar IRAS 16293-2422. An overview of a large molecular line survey from the Submillimeter Array. A&A 534:A100
16. Jørgensen JK, van der Wiel MHD, Coutens A, Lykke JM, Müller HSP, van Dishoeck EF, Calcutt H, Bjerkeli P, Bourke TL, Drozdovskaya MN, Favre C, Fayolle EC, Garrod RT, Jacobsen SK, Öberg KI, Persson MV, Wampfler SF (2016) The ALMA protostellar interferometric line survey (PILS). First results from an unbiased submillimeter wavelength line survey of the Class 0 protostellar binary IRAS 16293-2422 with ALMA. A&A 595:A117
17. Knude J, Hog E (1998) Interstellar reddening from the HIPPARCOS and TYCHO catalogues. I. Distances to nearby molecular clouds and star forming regions. A&A, 338:897–904
18. Kuan Y-J, Huang H-C, Charnley SB, Hirano N, Takakuwa S, Wilner DJ, Liu S-Y, Ohashi N, Bourke TL, Qi C, Zhang Q (2004) Organic molecules in low-mass protostellar hot cores: submillimeter imaging of IRAS 16293–2422. ApJ 616(1):L27–L30

19. Loison J-C, Halvick P, Bergeat A, Hickson KM, Wakelam V (2012) Review of OCS gas-phase reactions in dark cloud chemical models. MNRAS 421(2):1476–1484
20. Looney LW, Mundy LG, Welch WJ (2000) Unveiling the circumstellar envelope and disk: a subarcsecond survey of circumstellar structures. ApJ 529(1):477–498
21. Lundgren A (2012) ALMA Cycle 1 technical handbook
22. McElroy D, Walsh C, Markwick AJ, Cordiner MA, Smith K, Millar TJ (2013) The UMIST database for astrochemistry 2012. A&A 550:A36
23. Miura H, Yamamoto T, Nomura H, Nakamoto T, Tanaka KK, Tanaka H, Nagasawa M (2017) Comprehensive study of thermal desorption of grain-surface species by accretion shocks around protostars. ApJ 839(1):47
24. Mizuno A, Fukui Y, Iwata T, Nozawa S, Takano T (1990) A remarkable multilobe molecular outflow: rho ophiuchi east, associated with IRAS 16293–2422. ApJ 356:184
25. Müller HSP, Schlöder F, Stutzki J, Winnewisser G (2005) The cologne database for molecular spectroscopy, CDMS: a useful tool for astronomers and spectroscopists. J Molecular Struct 742(1-3):215–227
26. Mundy LG, Wootten A, Wilking BA, Blake GA, Sargent AI (1992) IRAS 16293–2422: a very young binary system? ApJ 385:306
27. Mundy LG, Wootten HA, Wilking BA (1990) The circumstellar structure of IRAS 16293-2422: C 18O, NH 3, and CO observations. ApJ 352:159
28. Ortiz-León GN, Loinard L, Kounkel MA, Dzib SA, Mioduszewski AJ, Rodríguez LF, Torres RM, González-Lópezlira RA, Pech G, Rivera JL, Hartmann L, Boden AF, II Evans NJ, Briceño C, Tobin JJ, Galli PAB, Gudehus D (2017) The Gould's belt distances survey (GOBELINS). I. Trigonometric parallax distances and depth of the ophiuchus complex. ApJ 834(2):141
29. Pickett HM, Poynter RL, Cohen EA, Delitsky ML, Pearson JC, Müller HSP (1998) Submillimeter, millimeter and microwave spectral line catalog. J Quant Spectro Radiative Trasfer 60(5):883–890
30. Pineda JE, Maury AJ, Fuller GA, Testi L, García-Appadoo D, Peck AB, Villard E, Corder SA, van Kempen TA, Turner JL, Tachihara K, Dent W (2012) The first ALMA view of IRAS 16293-2422. Direct detection of infall onto source B and high-resolution kinematics of source A. A&A 544:L7
31. Rao R, Girart JM, Marrone DP, Lai S-P, Schnee S (2009) IRAS 16293: A "Magnetic" Tale of two cores. ApJ 707(2):921–935
32. Sakai N, Oya Y, López-Sepulcre A, Watanabe Y, Sakai T, Hirota T, Aikawa Y, Ceccarelli C, Lefloch B, Caux E, Vastel C, Kahane C, Yamamoto S (2016) Subarcsecond analysis of the infalling-rotating envelope around the class I protostar IRAS 04365+2535. ApJ 820(2):L34
33. Sakai N, Oya Y, Sakai T, Watanabe Y, Hirota T, Ceccarelli C, Kahane C, Lopez-Sepulcre A, Lefloch B, Vastel C, Bottinelli S, Caux E, Coutens A, Aikawa Y, Takakuwa S, Ohashi N, Yen H-W, Yamamoto S (2014) A chemical view of protostellar-disk formation in L1527. ApJ 791(2):L38
34. Sakai N, Sakai T, Hirota T, Watanabe Y, Ceccarelli C, Kahane C, Bottinelli S, Caux E, Demyk K, Vastel C, Coutens A, Taquet V, Ohashi N, Takakuwa S, Yen H-W, Aikawa Y, Yamamoto S (2014) Change in the chemical composition of infalling gas forming a disk around a protostar. Nature 507(7490):78–80
35. Schöier FL, Jørgensen JK, van Dishoeck EF, Blake GA (2002) Does IRAS 16293–2422 have a hot core? Chemical inventory and abundance changes in its protostellar environment. A&A 390:1001–1021
36. Takakuwa S, Ohashi N, Bourke TL, Hirano N, Ho PT, Jørgensen JK, Kuan Y-J, Wilner DJ, Yeh SCC (2007) Arcsecond-resolution submillimeter HCN imaging of the binary protostar IRAS 16293-2422. ApJ 662(1):431–442
37. van der Tak FFS, Black JH, Schöier FL, Jansen DJ, van Dishoeck EF (2007) A computer program for fast non-LTE analysis of interstellar line spectra. With diagnostic plots to interpret observed line intensity ratios. A&A 468(2):627–635
38. van Dishoeck EF, Blake GA, Jansen DJ, Groesbeck TD (1995) Molecular abundances and low-mass star formation. II. Organic and deuterated species toward IRAS 16293-2422. ApJ 447:760

39. Wakelam V, Hersant F, Herpin F (2011) Sulfur chemistry: 1D modeling in massive dense cores. A&A 529:A112
40. Wootten A (1989) The duplicity of IRAS 16293–2422: a protobinary star? ApJ 337:858
41. Yeh SCC, Hirano N, Bourke TL, Ho PTP, Lee C-F, Ohashi N, Takakuwa S (2008) The CO molecular outflows of IRAS 16293-2422 probed by the submillimeter array. ApJ 675(1):454–463
42. Zapata LA, Loinard L, Rodríguez LF, Hernández-Hernández V, Takahashi S, Trejo A, Parise B (2013) ALMA 690 GHz observations of IRAS 16293–2422B: infall in a highly optically thick disk. ApJ 764(1):L14

Chapter 7
IRAS 16293–2422 Source B

7.1 Introduction

The existence of the infalling-rotating envelope and its centrifugal barrier have been confirmed in a prototypical hot corino source IRAS 16293−2422 Source A by analyzing its kinematic structure in the vicinity of the protostar (Chap. 6). Moreover, a drastic chemical change was found across the centrifugal barrier in this source; different molecular species trace different parts of the disk/envelope system. In Chaps. 4−6, sources with nearly edge-on configurations were investigated. Such chemical diagnostics will be a powerful tool to disentangle the infalling-rotating envelope and the Keplerian disk component even for sources with a face-on configuration, where the kinematic structure is difficult to be detected. Therefore, a similar analysis to the face-on source IRAS 16293−2422 Source B is conducted in this chapter. The usability of the chemical diagnostics is demonstrated, and it is confirmed whether the chemical change found in IRAS 16293−2422 Source A is seen in another hot corino source IRAS 16293−2422 Source B.

IRAS 16293−2422 Source B is one component of the binary system of IRAS 16293−2422 (d = 120 pc [8]) as well as IRAS 16293−2422 Source A, which have been investigated in Chap. 6. It is known to be rich in complex organic molecules (COMs) as well as Source A (e.g. [1, 5–7, 9, 15]). The disk/envelope system of Source B is reported to be nearly face-on, whereas Source A has a nearly edge-on geometry (Chap. 6; e.g. [15, 20]). Thus, observed line widths of molecular line emissions are narrower toward Source B than toward Source A. Furthermore, an inverse P-Cygni profile is reported toward Source B (e.g. [7, 15]), implying that there is infalling ambient gas in front of the protostar along the line of sight. In this chapter, the molecular distributions and the kinetic structures in this source are analyzed at a subarcsecond resolution, and the results are compared with those for Source A (Chap. 6).

This chapter has been published in Oya et al., 2018, ApJ, 854, 96.

Y. Oya, *A Few Tens au Scale Physical and Chemical Structures Around Young Low-Mass Protostars*, Springer Theses, https://doi.org/10.1007/978-981-19-1708-0_7

7.2 Observation

The analysis described in this chapter is based on the ALMA Cycle 1 archival data toward IRAS 16293−2422 used in Chap. 6 (see Appendix in Chap. 6 for the details). Spectral lines of OCS ($J = 19 - 18$), CH_3OH ($5_{1,5} - 4_{1,4}$; A^+), $HCOOCH_3$ ($20_{3,17} - 19_{3,16}$; A), and H_2CS ($7_{0,7} - 6_{0,6}$; $7_{2,5} - 6_{2,4}$; $7_{4,4} - 6_{4,3}$, $7_{4,3} - 6_{4,2}$) in the frequency range of 230 − 250 GHz were observed with the Band 6 receiver. As well, the ALMA Cycle 3 data is analyzed, which were carried out on 5 March 2016. The spectral line of SiO ($J = 6 - 5$; 260.518 GHz) was observed with the Band 6 receiver. 41 antennas were used in this observation, where the baseline length ranged from 17 to 636 m. The phase center of the observation was (α_{2000}, δ_{2000}) = ($16^h32^m22\overset{s}{.}87$, $-24°28'36\overset{''}{.}3$), and the primary beam width (FWHM) is 25″. The total on-source time was 16.38 minutes with typical system temperature ranging from 60 to 140 K. The backend correlator was tuned to a resolution of 122 kHz, corresponding to the velocity resolution of 0.14 km s^{-1} at 260 GHz, and a bandwidth of 58.6 MHz. The phase calibration was performed with J1625−2527 for every 7 minutes, and the bandpass calibration was with the quasar J1427−4206. The absolute flux density scale was derived from Titan. The data calibration was performed in the antenna-based manner, and the uncertainties of the observed intensities are expected to be less than 10% [16].

The spectral lines of OCS ($J = 19 - 18$), CH_3OH ($5_{1,5} - 4_{1,4}$; A^+), $HCOOCH_3$ ($20_{3,17} - 19_{3,16}$; A), H_2CS ($7_{0,7} - 6_{0,6}$; $7_{2,5} - 6_{2,4}$; $7_{4,4} - 6_{4,3}$, $7_{4,3} - 6_{4,2}$), and SiO ($J = 6 - 5$) are analyzed. Their rest frequencies, upper state energies, and intrinsic line strengths are summarized in Table 7.1. To obtain the continuum and spectral line images, the Brigg's weighting with the robustness parameter of 0.5 was employed. For the ALMA Cycle 1 data, self-calibration was applied by using the continuum emission. Self-calibration was not applied to the ALMA Cycle 3 data, including the SiO line, because the sensitivity of the continuum data is insufficient due to the limited number of line-free channels.

7.3 Distribution

Figure 7.1 shows the 1.2 mm continuum map. There are two intensity peaks corresponding to the two components of the binary (Source A, Source B). A weak emission is seen bridging between them. These features are consistent with the previous report [6]. The continuum distribution around Source B has an almost round shape, while that around Source A is slightly elongated along the NE-SW direction (see Chap. 6). These distributions are consistent with the geometries previously reported for the two sources; Source A and Source B are reported to have nearly edge-on and face-on configurations, respectively (e.g. [3, 17]). The sizes of the continuum emission deconvolved by the beam are evaluated by using the two-dimensional Gaussian fitting; ($1\overset{''}{.}138 \pm 0\overset{''}{.}007$) × ($0\overset{''}{.}583 \pm 0\overset{''}{.}004$) (P.A. $36\overset{\circ}{.}1 \pm 0\overset{\circ}{.}35$) for Source A,

Table 7.1 Parameters of the observed lines

Molecule	Transition	Frequency (GHz)	$E_u k_B^{-1}$ (K)	$S\mu^{2a}$ (D^2)	Beam Size
$OCS^{b,c}$	$J = 19 - 18$	231.0609934	111	9.72	0."65 × 0."52 (P.A. 84°.49)
$CH_3OH^{b,c}$	$5_{1,5} - 4_{1,4}$; A+	239.746253	49	3.89	0."53 × 0."46 (P.A. 71°.45)
$HCOOCH_3^{c,d}$	$20_{3,17} - 19_{3,16}$; A	250.25837	134	51.1	0."61 × 0."46 (P.A. 79°.40)
$H_2CS^{b,c}$	$7_{0,7} - 6_{0,6}$	240.2668724	46	19.0	0."53 × 0."46 (P.A. 73°.50)
$H_2CS^{b,c}$	$7_{2,5} - 6_{2,4}$	240.5490662	99	17.5	0."53 × 0."46 (P.A. 73°.67)
$H_2CS^{b,c,e}$	$7_{4,4} - 6_{4,3}$, $7_{4,3} - 6_{4,2}$	240.3321897	257	12.8, 12.8	0."53 × 0."46 (P.A. 73°.58)
$SiO^{b,f}$	$J = 6 - 5$	260.5180090	43.8	57.6	0."74 × 0."59 (P.A. 89°.09)

[a]Nuclear spin degeneracy is not included.
[b]Taken from CDMS [13].
[c]The ALMA Cycle 1 archival data. The self-calibration is employed.
[d]Taken from JPL [14].
[e]The rest frequencies of the two lines of H_2CS ($7_{4,4} - 6_{4,3}$; $7_{4,3} - 6_{4,2}$) are identical. They are denoted as $7_4 - 6_4$ in the text.
[f]The ALMA Cycle 3 data. The self-calibration is not employed

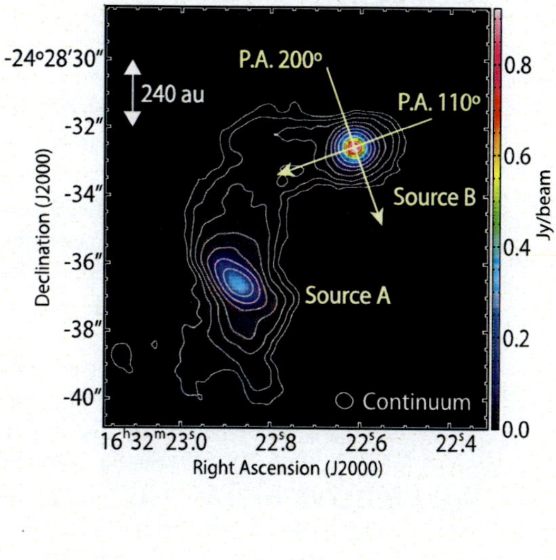

Fig. 7.1 The 1.2 mm continuum image with the beam size depicted in the bottom right corner. Contour levels are 10, 20, 40, 80, 160, 320, 640, 1280, and 2560σ, where the rms noise level is 0.3 mJy beam^{-1}

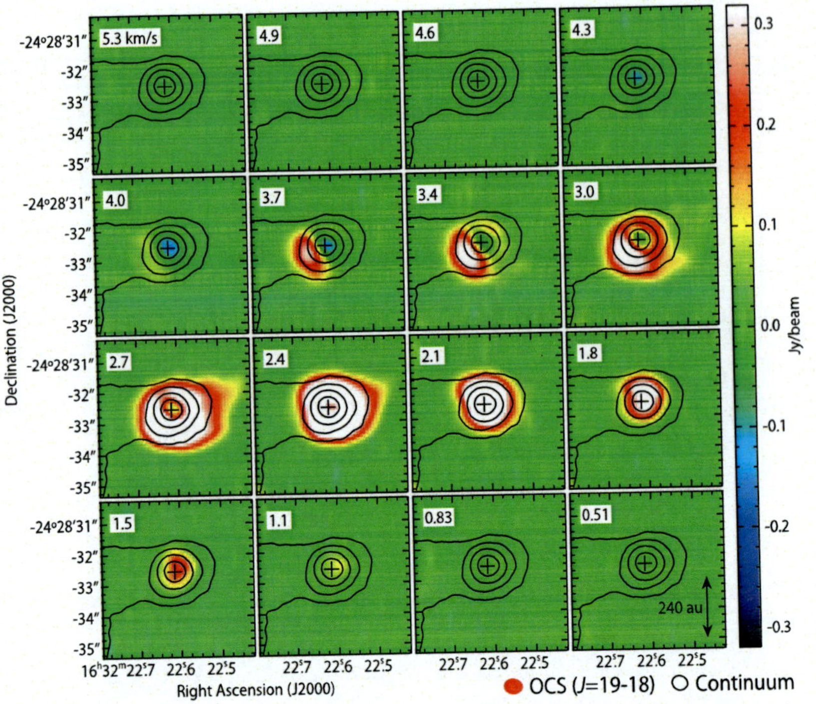

Fig. 7.2 Velocity channel maps of the OCS ($J = 19 - 18$) line. The number in the top left corner of each panel stands for the v_{LSR}. Contours represent the continuum map. Contour levels are 20, 80, 320, and 1280σ, where the rms noise level is 0.3 mJy beam^{-1}. The continuum peak position is represented by the black crosses

and ($0''.429 \pm 0''.003$) × ($0''.357 \pm 0''.003$) (P.A. $139^\circ.0 \pm 1^\circ.6$) for Source B. The peak flux densities are (314.0 ± 1.4) and (899.9 ± 1.5) mJy beam^{-1} for Source A and Source B, respectively.

The velocity channel maps of the OCS ($J = 19 - 18$) line are shown in Fig. 7.2. Absorption features toward the continuum peak position are seen in the maps for v_{LSR} from 3.4 to 4.3 km s^{-1}. Since the systemic velocity of IRAS 16293−2422 Source B is around 3 km s^{-1} [1], these absorption features are red-shifted. They are most naturally interpreted as the inverse P-Cygni profile, as previously reported [7, 15, 20]. At the systemic velocity, the OCS distribution is extended at a 3″ (∼400 au) scale in diameter around the continuum peak position, although it would suffer from the resolving-out effect with the largest recoverable scale of ∼14″ in this observation.

Figures 7.3, 7.4, and 7.5 show the velocity channel maps of the CH_3OH ($5_{1,5} - 4_{1,4}$; A$^+$), $HCOOCH_3$ ($20_{3,17} - 19_{3,16}$; A), and H_2CS ($7_{0,7} - 6_{0,6}$) lines, respectively. As in the OCS case, absorption features toward the continuum peak position are found for all these three molecular lines in their red-shifted components ($v_{\mathrm{LSR}} = 3 - 5$ km s^{-1}). The distributions of CH_3OH, $HCOOCH_3$, and H_2CS are more compact

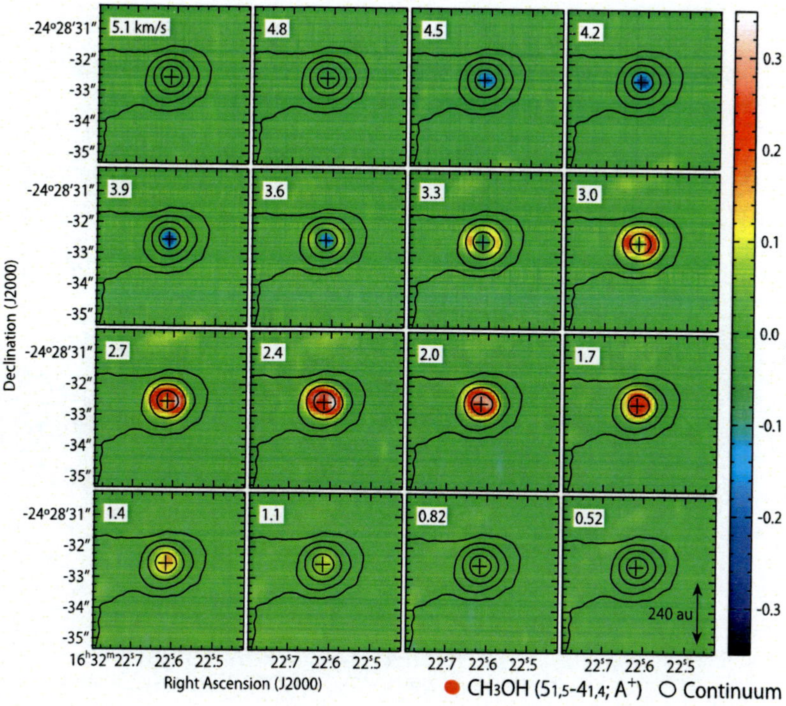

Fig. 7.3 Same as Fig. 7.2, but for the CH_3OH ($5_{1,5}-4_{1,4}$; A^+) line

than that of OCS. Among them, the H_2CS distribution is larger than the CH_3OH and $HCOOCH_3$ distributions. The H_2CS emission is slightly extended toward the western side of the continuum peak.

These results for OCS, CH_3OH, $HCOOCH_3$, and H_2CS show small-scale chemical differentiation in the vicinity of the protostar in Source B. It is similar to that found in Source A (Chap. 6); the CH_3OH and $HCOOCH_3$ emissions are concentrated around the centrifugal barrier with the radius of 50 au, while the OCS and H_2CS emissions show more extended distributions.

Figure 7.6 shows the integrated intensity maps of the six molecular lines of the above four molecular species. Except for the $HCOOCH_3$ line, the intensity distributions show a ring-like structure around the protostar; the line intensities are weaker toward the continuum peak position than the surrounding positions with a radius of $0''.25$ (~30 au). This is caused by the contribution of the inverse P-Cygni profile toward the continuum peak position found in the channel maps (Figs. 7.2, 7.3, and 7.5).

As seen in the channel maps of the molecular lines, difference in the size of their distribution is confirmed in their integrated intensity maps. Specifically, the OCS emission shows the most extended distribution over $1''$ (~120 au) around the

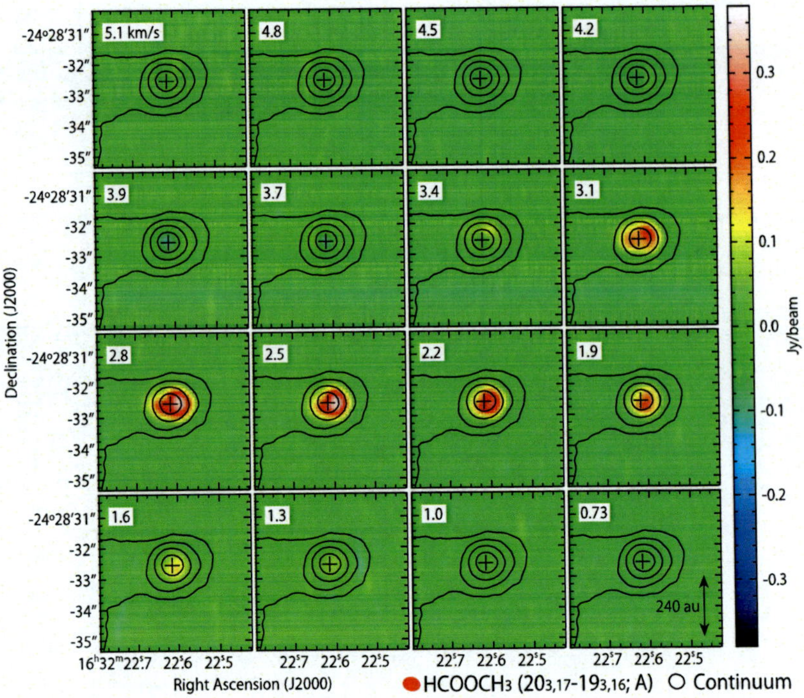

Fig. 7.4 Same as Fig. 7.2, but for the $HCOOCH_3$ ($20_{3,17} - 19_{3,16}$; A) line

protostar. On the contrary, the CH_3OH and $HCOOCH_3$ distributions are concentrated around the protostar with a radius of 0.″6 (~70 au). The H_2CS ($7_{0,7} - 6_{0,6}$) distribution is slightly more extended than the CH_3OH and $HCOOCH_3$ distributions, but is not so extended as the OCS distribution. The higher excitation lines of H_2CS ($7_{2,5} - 6_{2,4}$; $7_{4,4} - 6_{4,3}$, $7_{4,3} - 6_{4,2}$) show distributions as compact as the CH_3OH and $HCOOCH_3$ lines.

These line emissions have the velocity widths as narrow as 3 km s^{-1} at FWHM (full width at half maximum) toward Source B, while they are typically wider than 5 km s^{-1} toward Source A. The observed narrow velocity widths likely originate from the nearly face-on geometry of the disk/envelope system of Source B. Nevertheless, a small velocity gradient can be recognized in the channel maps of the OCS line (Fig. 7.2). At the blue-shifted velocity (v_{LSR} ~1.5 km s^{-1}), the OCS distribution has a slight offset from the continuum peak position toward the northwestern direction. On the contrary, it has a slight offset toward the southeastern direction at the red-shifted velocity (v_{LSR} ~4 km s^{-1}). This trend can be confirmed in Fig. 7.7, where the integrated intensity maps of the high velocity-shift components are shown. Although the integrated intensity of the red-shifted component is affected by the absorption toward the continuum peak, the red-shifted component of the OCS line, the con-

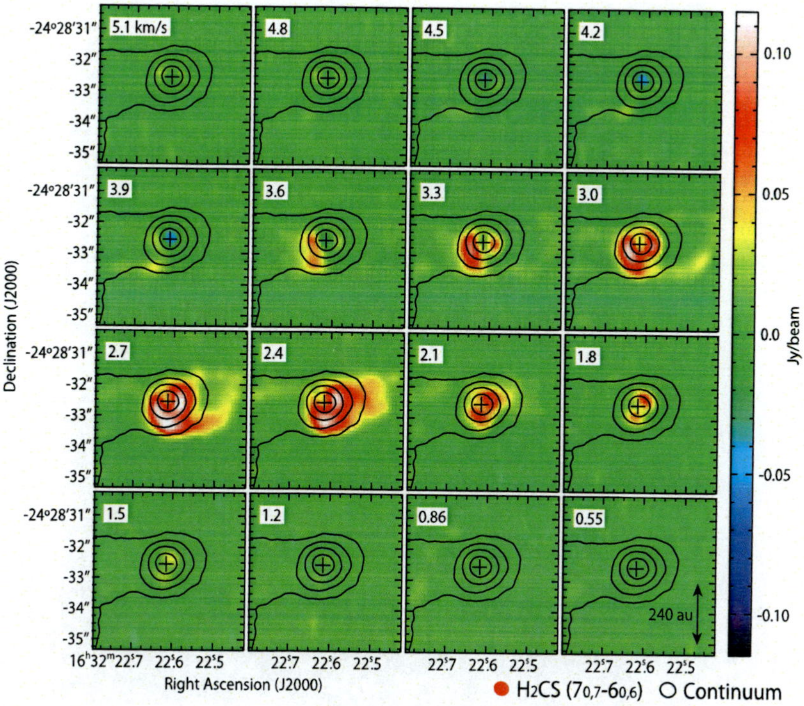

Fig. 7.5 Same as Fig. 7.2, but for the H_2CS ($7_{0,7} - 6_{0,6}$) line

tinuum peak position, and the blue-shifted component of the OCS line seem to be alined on a common line along the southeast-northwest direction with a position angle (P.A.) of about 110°. This velocity gradient likely suggests the rotation motion in the disk/envelope system slightly inclined from the face-on geometry.

7.4 Kinematic Structure

7.4.1 Observed Features

The disk/envelope system of IRAS 16293−2422 Source B is thought to have a nearly face-on geometry. Since a hint of a rotation motion is seen (Sect. 7.3), the disk/envelope system is expected to be slightly inclined. To investigate the velocity structure of the gas around the protostar, the position-velocity (PV) diagrams of the OCS, H_2CS, CH_3OH, and $HCOOCH_3$ lines are prepared (Figs. 7.8, 7.9, 7.10 and 7.11).

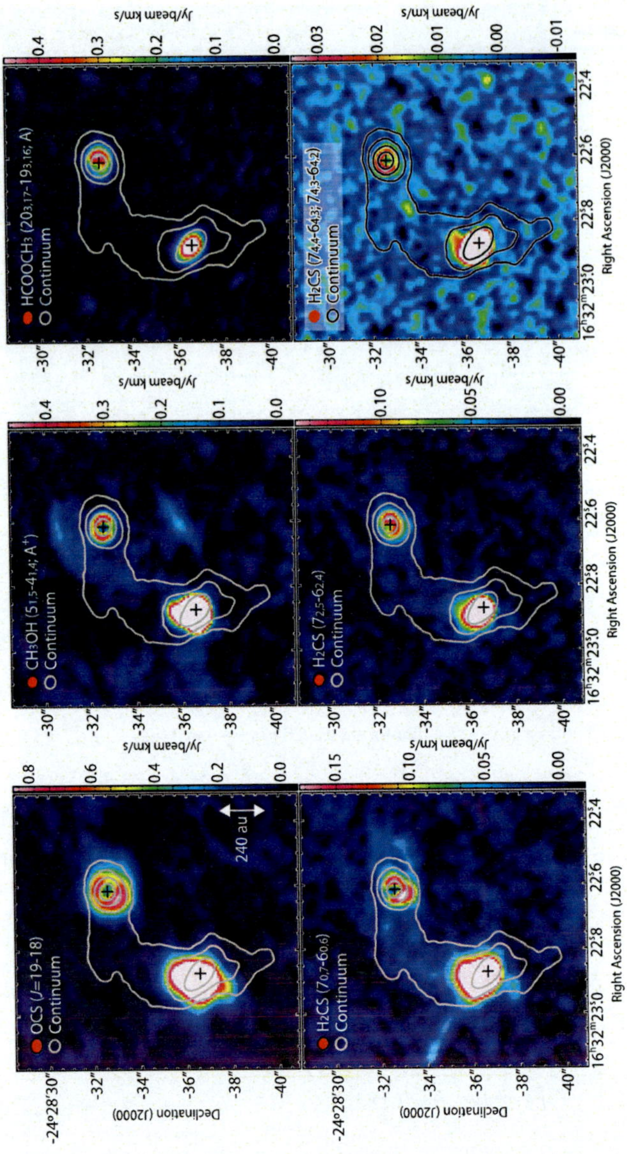

Fig. 7.6 Integrated intensity maps of the OCS ($J = 19 - 18$), CH_3OH ($5_{1,5} - 4_{1,4}$; A^+), $HCOOCH_3$ ($20_{3,17} - 19_{3,16}$; A), and H_2CS ($7_{0,7} - 6_{0,6}$; $7_{2,5} - 6_{2,4}$; $7_{4,4} - 6_{4,3}$, $7_{4,3} - 6_{4,2}$) lines associated with Source B (color). The velocity-shift range for the integration is $\pm 2\ \mathrm{km\ s^{-1}}$. Contours represent the continuum map. Contour levels are as the same as those in Fig. 7.2. Black crosses represent the continuum peak position

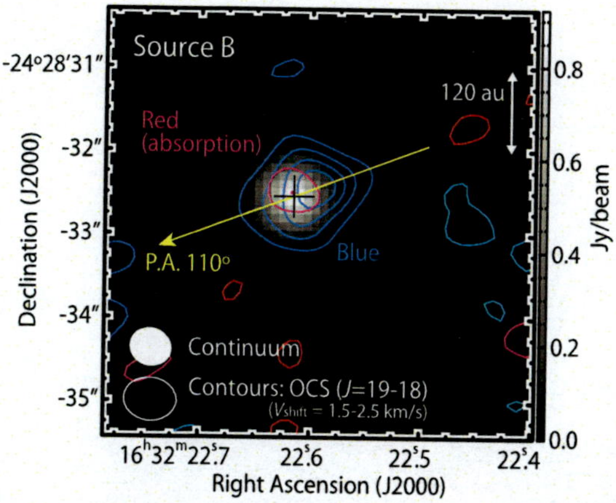

Fig. 7.7 Integrated intensity maps of high velocity-shift components of the OCS ($J = 19 - 18$; contours) line overlaid on the continuum map (gray scale). Red and magenta contours represent the red-shifted components with the velocity range from 4.4 to 5.4 km s^{-1}. Blue and cyan ones represent the blue-shifted ones with the velocity range from 0.4 to 1.4 km s^{-1}. Magenta and cyan contours represent the negative values. Contour levels are every 3σ, where the rms noise level is 5 mJy beam^{-1} km s^{-1}. Black cross represents the continuum peak position

The inverse P-Cygni profiles toward the protostar are confirmed in the PV diagrams. In Fig. 7.8a, the OCS emission shows a slight velocity gradient along the P.A. of 110°, although it is heavily contaminated with the absorption feature toward the protostellar position. Specifically, the OCS emission tends to have a more blue-shifted velocity on the northwestern side than on the southeastern side. The intensity peak position in this diagram is slightly offset from the protostellar position. As well, the diagrams prepared along the lines with other P.A.s show the absorption features toward the protostellar position. Moreover, two intensity peaks are found with offsets from the protostellar position in the diagrams for the P.A. of 140°, 170°, 200°, and 230°. For instance, in the diagram for the P.A. of 200°, the peak intensity (~0.6 Jy beam^{-1}) is about 1.5 times higher than the intensity toward the protostellar position (~0.4 Jy beam^{-1}).

As well as the OCS case, the PV diagrams of the H_2CS ($7_{0,7} - 6_{0,6}$) line show a slight velocity gradient along the P.A. of 110° (Figure 7.9a). In this diagram, the absorption feature toward the protostellar position is confirmed in the red-shifted components. There is an additional absorption feature seen at the velocity higher than 5 km s^{-1}; this feature is likely a contamination by an unidentified line. The diagrams of the H_2CS lines show two intensity peaks, except for that for the P.A. of 200°, as in the case of OCS. The intensity dips toward the protostellar position are seen clearer in the H_2CS emission than in the OCS emission. For instance, the peak intensity (~0.12 Jy beam^{-1}) is twice higher than the intensity toward the protostellar potision (0.05 Jy beam^{-1}) in the diagram for the P.A. of 110°. Asymmetry is found for the peak intensities in the PV diagrams of the H_2CS lines; the peak intensity is

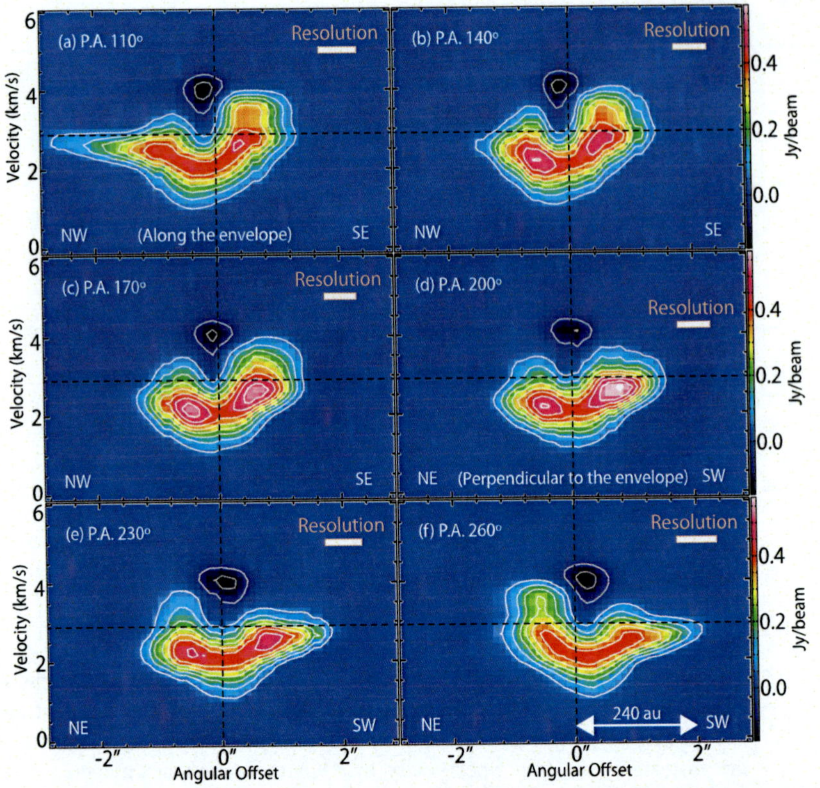

Fig. 7.8 Position-velocity diagrams of the OCS ($J = 19 - 18$) line. Diagrams are prepared along the lines centered at the protostellar position, whose P.A.s are taken for every 30° from 110° as depicted in the top left corners of the panels. Contour levels are from -40σ and every 20σ, except for 0σ, where the rms noise level is 3.2 mJy beam^{-1}. Rectangle in each panel represents the spatial and velocity resolutions

higher on the southern side of the protostar than on the northern side, except for the diagram for the P.A. of 110°. This trend may be caused by asymmetry in the gas distribution.

On the contrary, such a velocity gradient is not seen in the PV diagrams of the CH_3OH line (Fig. 7.10). In the PV diagram for the P.A. of 110° (Fig. 7.10a), the CH_3OH emission has a distribution as compact as 1.″5. Its maximum velocity-shift is 2 km s^{-1} from the systemic velocity, which is comparable to that for the OCS emission. The absorption in the red-shifted component is confirmed toward the continuum peak position. The size of the absorption feature is less than 1″, which is slightly smaller than that of the size of the CH_3OH distribution. The PV diagrams of the CH_3OH line prepared for six P.A.s are essentially similar to one another in the sizes of the absorption features and the distributions of the emission, although they show some

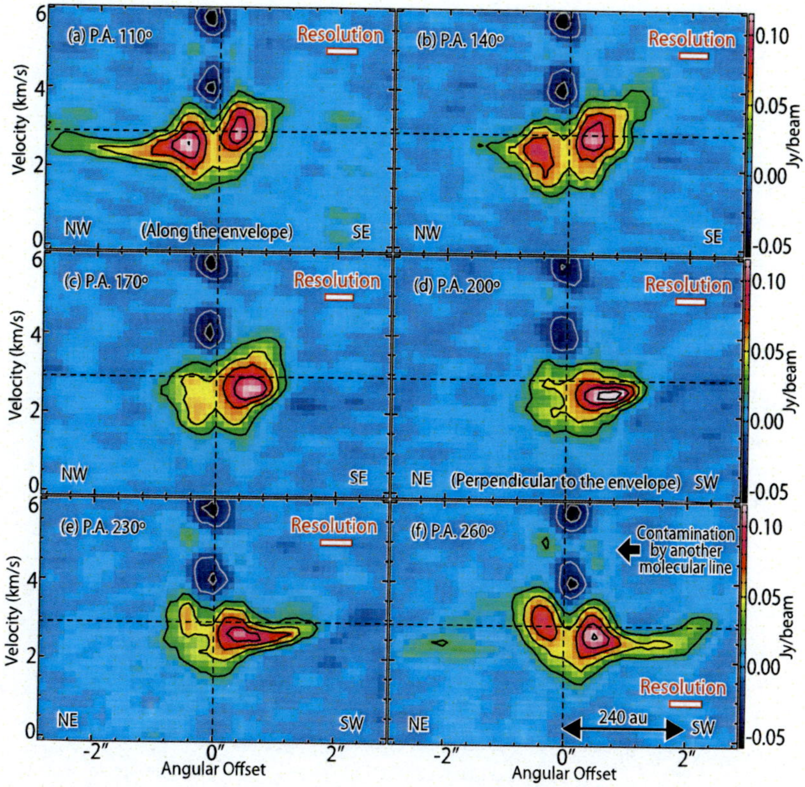

Fig. 7.9 Same as Fig. 7.8, but for the H_2CS ($7_{0,7} - 6_{0,6}$) line. Contour levels are from -20σ and every 10σ, except for 0σ, where the rms noise level is 2.3 mJy beam^{-1}. Contaminations by an unidentified line are seen at a velocity offset of $+2$ km s^{-1}

asymmetry with respect to the protostellar position. For the PV diagrams for the P.A. from 140° to 200°, where the northern component is intense, the asymmetry in the CH_3OH intensity seems to be anti-correlated to that seen in the H_2CS line. The PV diagrams of the $HCOOCH_3$ line are shown in Fig. 7.11a, b. As well as the CH_3OH line, the $HCOOCH_3$ line shows a compact distribution without a clear velocity gradient.

7.4.2 Comparison of Molecular Distribution with the Source A Case

Since the rotation motion around Source B is marginally revealed in the OCS and H_2CS lines, their kinematic structures are investigated in more detail. As for Source

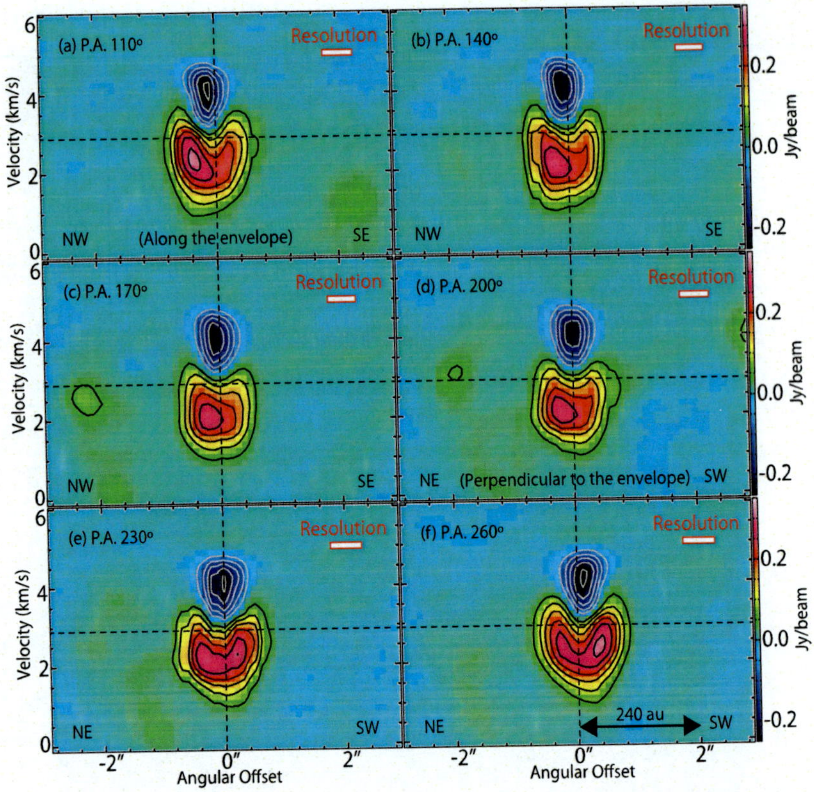

Fig. 7.10 Same as Fig. 7.8, but for the CH_3OH ($5_{1,5} - 4_{1,4}$; A^+) line. Contour levels are from -100σ and every 20σ, except for 0σ, where the rms noise level is 2.5 mJy beam^{-1}

A, the kinematic structure of the infalling-rotating envelope at a 100 au scale was successfully reproduced by a ballistic model (Chap. 6). Hence, the advantage of this knowledge is taken in the analysis of the Source B data.

In Source A, its infalling-rotating envelope is traced by the OCS emission, while its centrifugal barrier is highlighted by the COM emissions (Sect. 7.1; Chap. 6). This is exactly the case in Source B. Hence, the physical structure of the gas around Source B can be explored in a similar way by using the chemical differentiation. If the OCS emission traces the infalling-rotating envelope, the velocity gradient detected in its PV diagrams (Fig. 7.8) can be attributed to a combination of the infall and rotation motions in the envelope. Since CH_3OH and $HCOOCH_3$ seem to be evaporated from dust grains inside the centrifugal barrier in Source A, their distributions in Source B would be comparable to the size of the centrifugal barrier.

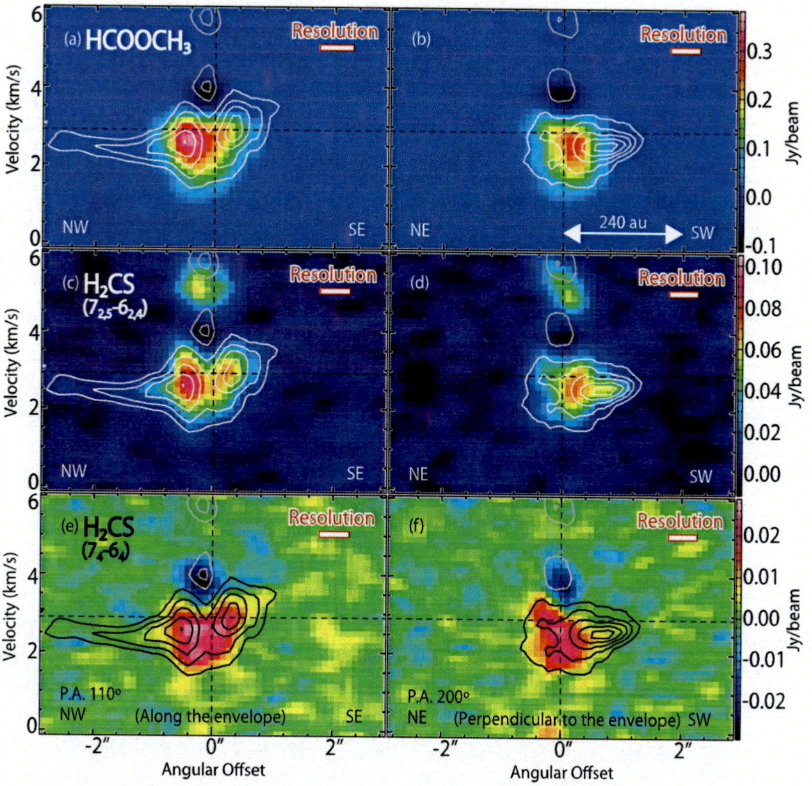

Fig. 7.11 Position-velocity diagrams of the $HCOOCH_3$ ($20_{3,17} - 19_{3,16}$; A; **a**, **b**) and two H_2CS ($7_{2,5} - 6_{2,4}$ for panels **c**, **d**; $7_{4,4} - 6_{4,3}$, $7_{4,3} - 6_{4,2}$ for panels **e**, **f**) lines. Position axes are as the same as those in Fig. 7.8a, d, which are taken for the P.A.s of 110° and 200°. Contours in each panel represent the corresponding PV diagram of the H_2CS ($7_{0,7} - 6_{0,6}$) line in Fig. 7.9a, d

Meanwhile, the H_2CS emission shows a different behavior in Source B from that in Source A. It traces the disk component with high velocity-shifts inside the centrifugal barrier in Source A. In contrast, the H_2CS emission does not apparently show such components in Source B. Its PV diagram for the P.A. of 110° show double-peaked structures with an intensity dip toward the protostellar position (Fig. 7.9a). This dip cannot be attributed to the effect of the high dust opacity in this source. This is because emissions of other molecular lines are indeed detected toward the protostellar position, particularly their blue-shifted components: the CH_3OH, $HCOOCH_3$, and H_2CS ($7_{2,5} - 6_{2,4}$; $7_{4,4} - 6_{4,3}$, $7_{4,3} - 6_{4,2}$) lines show compact emissions there in their PV diagrams (Figs. 7.10, 7.11). Furthermore, the dip feature is not due to the absorption by the foreground gas either, because those in the blue-shifted components have the velocity shift as high as -1 km s^{-1} from the systemic velocity. Hence, the intensity dip means that H_2CS would be deficient in the closest vicinity of the

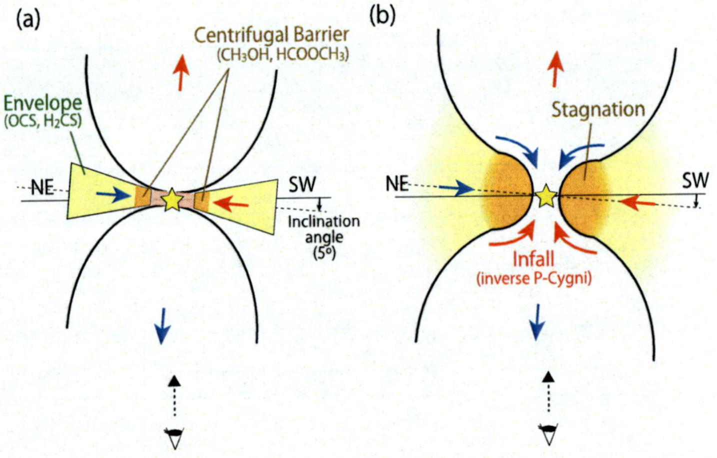

Fig. 7.12 Schematic illustration of the disk/envelope system in IRAS 16293−2422 Source B. In panel (**a**), the disk/envelope system has a thin structure. On the contrary, the gas is stagnated in panel (**b**), and has an infalling motion around the protostar

protostar. The two intensity peaks in the H_2CS emission (Fig. 7.9a) just correspond to the positions within which the CH_3OH and $HCOOCH_3$ emission appear (Figs. 7.10a, 7.11a). Considering that CH_3OH and $HCOOCH_3$ are concentrated around the centrifugal barrier in Source A, the intensity peaks in the H_2CS emission would represent the centrifugal barrier. Although the OCS emission does not clearly show such a double-peaked structure for the P.A. of 110°, its intensity peak is shifted from the protostellar position and almost coincides one of the intensity peaks in the H_2CS emission. Therefore, the two intensity peaks in the PV diagram of the H_2CS emission for the P.A. of 110° most likely represent the positions of the centrifugal barrier in Source B.

The configuration of the disk/envelope system in Source B discussed above is schematically shown in Fig. 7.12a. Furthermore, the outflow structure supports this configuration, as described later (Sect. 7.6).

7.5 Modelling

7.5.1 *Infalling-Rotating Envelope Model*

In this section, the velocity gradient traced by the OCS ($J = 19 - 18$) and H_2CS ($7_{0,7} - 6_{0,6}$) lines are investigated by using the infalling-rotating envelope model (Chap. 3). Specifically, it is clearly shown as the double-peaked structure in the PV diagram of the H_2CS line prepared for the P.A. of 110°, which is expected to be along

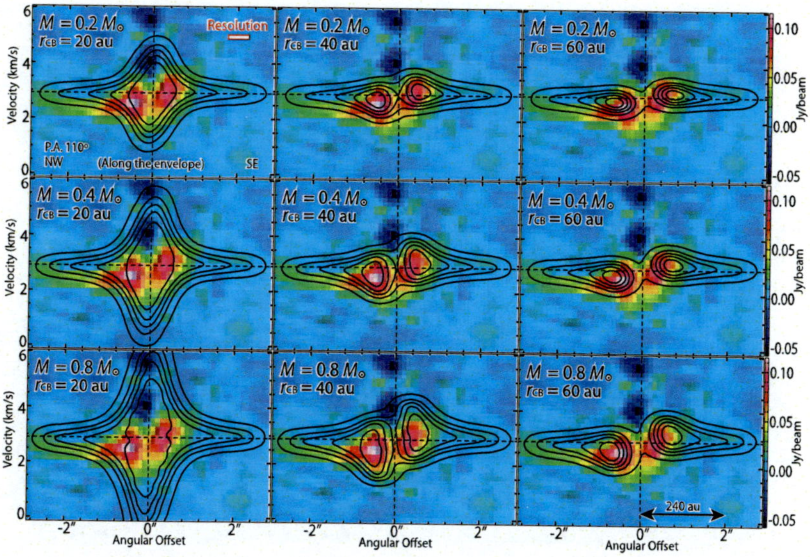

Fig. 7.13 Position-velocity diagram of the H_2CS ($7_{0,7} - 6_{0,6}$; color) prepared along the disk/envelope direction (P.A. 110°). Color maps are the same as that in Fig. 7.9a. The results of the infalling-rotating envelope models are represented by black contours. The parameters for the models are varied among the panels; M is 0.2, 0.4, or 0.8 $M_{\odot}$, and $r_{\rm CB}$ is 20, 40, or 60 au, where i is assumed to be 5°. Contour levels are every 20% from 5% of the peak intensity in each panel

the mid-plane of the disk/envelope system. Thus, the parameters for the model are evaluated by comparing this PV diagram in observation with the corresponding one simulated by the model.

Examples of the infalling-rotating envelope models are superposed on the PV diagrams of the H_2CS ($7_{0,7} - 6_{0,6}$) line in Figs. 7.13 and 7.14, where the position axes are prepared for the disk/envelope direction (P.A. 110°) and the direction perpendicular to it (P.A. 200°), respectively. The systemic velocity ($v_{\rm sys}$) is assumed to be 2.9 km s^{-1}. The main parameters of the infalling-rotating envelope model are the protostellar mass (M) and the radius of the centrifugal barrier ($r_{\rm CB}$). To demonstrate how sensitive the modeled PV diagrams are to these parameters, model calculations are conducted for various sets of the parameters. The envelope models are assumed to have a constant thickness of 50 au and an outer radius of 300 au with an inclination angle of 5° (0° for a face-on configuration). The model images are convolved with the beam size in the observation and the intrinsic line width of 0.5 km s^{-1}. In this model analysis, the density profile is assumed to be proportional to the distance to the protostar ($\propto r^{-1.5}$), and the molecular abundance is to be constant, where the effects of the optical depth, excitation, and the temperature gradient are not taken into account. The peak intensity of the OCS line is ~0.6 Jy beam^{-1}, which corresponds to the brightness temperature of 40 K. Thus, the OCS line seems to be optically thin with the comparison with the gas kinetic temperature derived below (see Sect. 7.7).

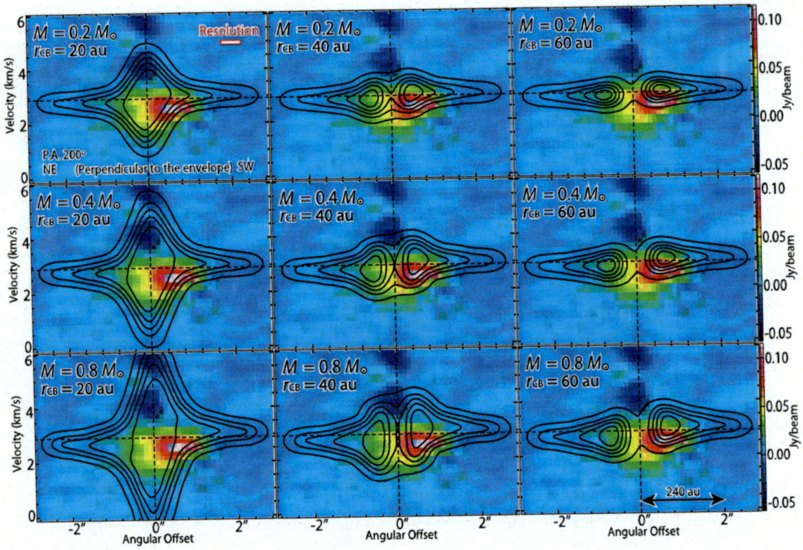

Fig. 7.14 Same as Fig. 7.13, but along the direction perpendicular to the disk/envelope direction (P.A. 200°). Color maps are the same as that in Fig. 7.9d

In these observations, the other molecular lines are weaker than the OCS line, and hence, they can also be assumed to be optically thin.

Needless to say, the absorption feature in the red-shifted component toward the protostar cannot be explained by this simple infalling-rotating envelope model, because it does not consider any radiative transfer effect (Chap. 3). However, the velocity gradient detected along the disk/envelope direction (P.A. 110°) is reasonably explained by the rotation motion in the model (Fig. 7.13). Especially, the models with r_{CB} of 40 au well reproduce the observation except for the absorption feature. The model is not very sensitive to the protostellar mass (M), and hence, it is weakly constrained around 0.4 $M_\odot$ within a factor of 2 as far as the inclination angle of 5° is assumed. Meanwhile, r_{CB} strongly affects the model results. Specifically, two intensity peaks in the PV diagrams correspond to the centrifugal barrier. With r_{CB} of 60 au, the distances between the peaks in model results are larger than that observed. With r_{CB} of 20 au, the emission in models are not spatially resolved in the vicinity of the protostar with the beam size for the H_2CS ($7_{0,7} - 6_{0,6}$) line, and thus, models cannot reproduce the observed double-peaked structure. Hence, the reasonable range for r_{CB} can be evaluated to be from 30 to 50 au. As well, the models with r_{CB} of 40 au indeed show a better agreement with the observed PV diagram prepared along the line perpendicular to the disk/envelope direction (P.A. 200°; Fig. 7.14), although some asymmetry is seen in the observation.

The infalling-rotating envelope model results are also compared with the OCS ($J = 19 - 18$) and CH_3OH ($5_{1,5} - 4_{1,4}$; A^+) lines in Fig. 7.15, where M of 0.4

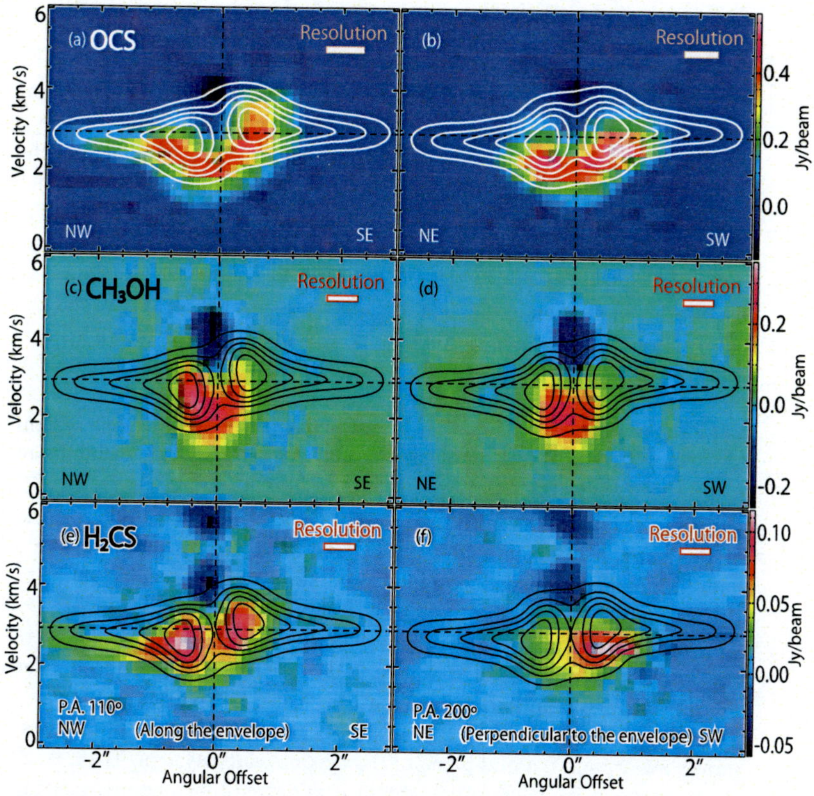

Fig. 7.15 Position-velocity diagrams of the OCS ($J = 19 - 18$; **a**, **b**), CH_3OH ($5_{1,5} - 4_{1,4}$; A^+; **c**, **d**), and H_2CS ($7_{0,7} - 6_{0,6}$; **e**, **f**) lines. Position axes are taken along the disk/envelope direction (P.A. 110°; **a**, **c**, **e**) and the direction perpendicular to it (P.A. 200°; **b**, **d**, **f**). Color maps are the same as those in panels (a) and (d) in Figs. 7.8, 7.9, and 7.10. Contours represent the results of the infalling-rotating envelope model with M of 0.4 $M_\odot$, r_{CB} of 40 au, and i of 5°. Contour levels are every 20% from 5% of the peak intensity in each panel

$M_\odot$ and r_{CB} of 40 au are employed as representative values. Strictly speaking, a double-peaked structure is not clear in the OCS line in comparison with the H_2CS ($7_{0,7} - 6_{0,6}$) line. Nevertheless, the model seems to explain the basic feature observed in the OCS line, considering that the absorption feature is not taken into account in the model. On the contrary, the model obtained from the above analysis does not explain the kinematic structure traced by the CH_3OH ($5_{1,5} - 4_{1,4}$; A^+) line. In the observational result, the emitting region of the CH_3OH ($5_{1,5} - 4_{1,4}$; A^+) line is likely concentrated around the centrifugal barrier and/or inside it. This is consistent with the case for IRAS 16293−2422 Source A (Chap. 6); the CH_3OH ($11_{0,11} - 10_{1,10}$; A^+) line mainly highlights the centrifugal barrier of Source A and/or inside it, namely the disk component.

7.5.2 Origin of the Inverse P-Cygni Profile

As discussed above, the basic features of the observed kinematic structures in the molecular emissions are reasonably explained by the infalling-rotating envelope model to some extent. However, the origin of the inverse P-Cygni profile is still remained as an important problem. Inverse P-Cygni profiles suggest an infall motion along the line of sight. Since the disk/envelope system of IRAS 16293−2422 Source B has a nearly face-on configuration, the outflow lobes are also expected to exist along the line of sight as described in Sect. 7.6. This situation, where the infall motion and the outflow motion along the line of sight coexist in the vicinity of the protostar, is hardly possible, as far as a thin disk structure is considered as shown in Fig. 7.12a. At least, a thick disk/envelope structure would be required so that substantial amount of infalling gas exists near the protostar. In fact, the simulations of outflow show such an infalling gas perpendicular to the mid-plane of the disk/envelope system near the launching point of outflows [11]. Even in this case, there remains the following difficulty. The absorption feature has a velocity shift of at most 2 km s^{-1}. Assuming the free fall motion from an infinite point, this velocity shift corresponds to the distance of 180 au to the protostar with a mass of 0.4 $M_{\odot}$. This distance seems to contradict with the compact distribution of the COM lines (∼50 au) accompanying the inverse P-Cygni profile.

One possible idea to solve this problem is the stagnation of the accreting gas. For instance, the protostellar source L1527 is reported that the thickness of its disk/envelope system is broadened at the centrifugal barrier and inward of it due to the stagnant gas [18]. At the centrifugal barrier, a part of the stagnant gas once moves away from the mid-plane of the disk/envelope system. Then, it falls toward the protostar if its angular momentum is extracted by some mechanisms, such as disk winds and/or low-velocity outflows. Such an infalling gas may cause the inverse P-Cygni profile. This situation is illustrated in Fig. 7.12b. It has previously been thought that the infalling gas in the ambient envelope causes the inverse P-Cygni profile [15]. However, the velocity shift of the absorbing gas (∼2 km s^{-1}) seems to be too large for the ambient component.

With this in mind, a free fall motion in the vicinity of the protostar is incorporated in the model. A spherical clump with a radius of the centrifugal barrier ($r_{CB} = 40$ au) is assumed for the gas distribution around the protostar. Figure 7.16 shows the results of this modified model. In this model, the gas is assumed to be once stagnated at the distance of r_{CB} to the protostar to make a spherical clump, and then, it falls toward the protostar under its gravity without any initial velocity. This very simplified picture can explain the kinematic structure at a 40 au scale around the protostar, including the inverse P-Cygni profile. In this case, the gas having the infall velocity of 2 km s^{-1}, which shows the inverse P-Cygni profile, is located at 33 au from the protostar. However, it should be emphasized that this model is just a simplified hypothetical one accounting for the hypothetical picture described above. To verify its validity, the gas structure in the vicinity of the protostar is required to be resolved in further observations.

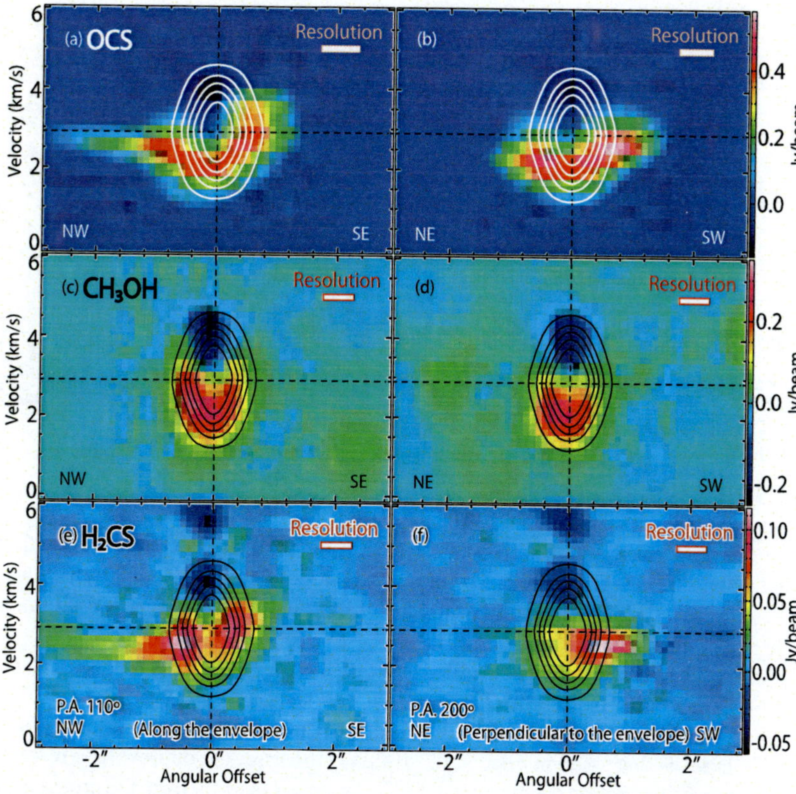

Fig. 7.16 Position-velocity diagrams of the OCS ($J = 19 - 18$; **a**, **b**), CH_3OH ($5_{1,5} - 4_{1,4}$; A^+; **c**, **d**), and H_2CS ($7_{0,7} - 6_{0,6}$; **e**, **f**) lines. Color maps are the same as those in Fig. 7.15, or the corresponding panels in Figs. 7.8, 7.9, and 7.10. Contours represent the model results with infall motion from the centrifugal barrier, where the accreting gas is stagnated once. The parameters for the model are M of 0.4 $M_{\odot}$, r_{CB} of 40 au, and i of 5°. The molecular line emission is assumed to come from the spherical region with the radius from 35 to 40 au around the protostar. Contour levels are every 20% from 5% of the peak intensity in each panel

7.6 Outflow

The SiO ($J = 6 - 5$) data is also analyzed. It is often employed as a shock tracer. The integrated intensity maps of its high velocity-shift components are shown in Fig. 7.17. The blue-shifted and red-shifted components are spatially overlapped with each other even at a 300 au scale, and moreover, both of them show a shell-like feature surrounding the protostellar position. This feature implies a nearly pole-on geometry of the outflow. Namely, the outflow cavity is looked at as a shell-like feature. Nevertheless, the blue-shifted components likely tend to extend toward the northeastern direction from the protostar. Meanwhile, the red-shifted emission is

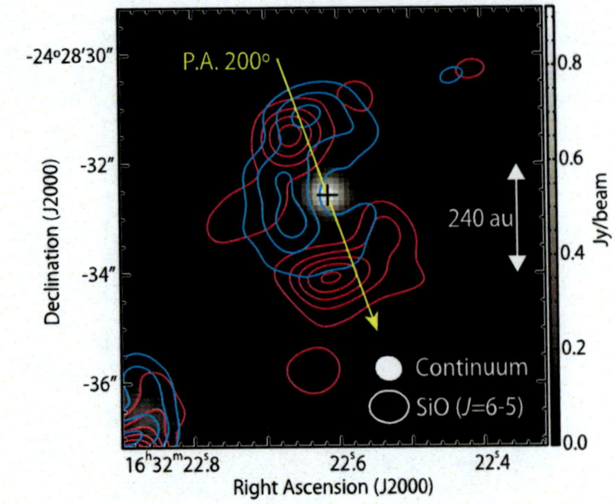

Fig. 7.17 Integrated intensity maps of the high velocity-shift components of the SiO ($J = 6 - 5$) line overlaid on the continuum map (gray scale). The velocity range for the integration is −0.2 to 1.7 km s^{-1} for the cyan contours, and from 3.8 to 5.8 km s^{-1} for the magenta ones. Contour levels are every 10 σ, where the rms noise level is 20 mJy $beam^{-1}$ km s^{-1}. Black cross represents the continuum peak position

more intense on the southwestern side of the protostar in comparison with the blue-shifted one. Therefore, the outflow axis might be inclined slightly as illustrated in Fig. 7.12. In this situation, the blue-shifted outflow lobe overlaps on the protostar along the line of sight.

The inverse P-Cygni profile is seen in the OCS, CH_3OH, $HCOOCH_3$, and H_2CS lines toward the protostellar position. The situation illustrated in Fig. 7.12b allows the coexistence of the infalling gas toward the protostar and the blue-shifted outflow lobe, as discussed in Sect. 7.5.2. If this is the case, the outflow lobe responsible for the SiO emission is not expected to be launched directly from the protostar but could be from the inner part of the disk/envelope system, possibly near the centrifugal barrier. In fact, such a situation is hinted at by the kinematic structure traced by the SiO emission. The bipolar outflow lobes are shown in the SiO emission to be along the P.A. of 200° (Fig. 7.18). Both the blue-shifted and red-shifted lobes are detected on both the northeastern and southwestern sides of the protostar, which is consistent with the configuration illustrated in Fig. 7.12b. If the outflowing gas is accelerated as it propagates away from the protostar (Chap. 3; e.g. [10]), the components near the systemic velocity likely come from the launching point of the outflow. The SiO emission at the systemic velocity is not detected toward the protostellar position but at the position with an offset of ∼0.″5 from it. This result implies that the outflow of this source has its launching point at a position offset from the protostar (Fig. 7.12b). Furthermore, this component in the SiO emission likely appears close to the peak positions of the OCS and H_2CS emissions (Fig. 7.18). This feature may suggest that the outflow could be launched around the centrifugal barrier traced by the OCS and H_2CS emissions.

Large-scale outflows blowing out from this binary system have extensively been studied (e.g. [4, 12]). The direction of the outflow found in this study is different

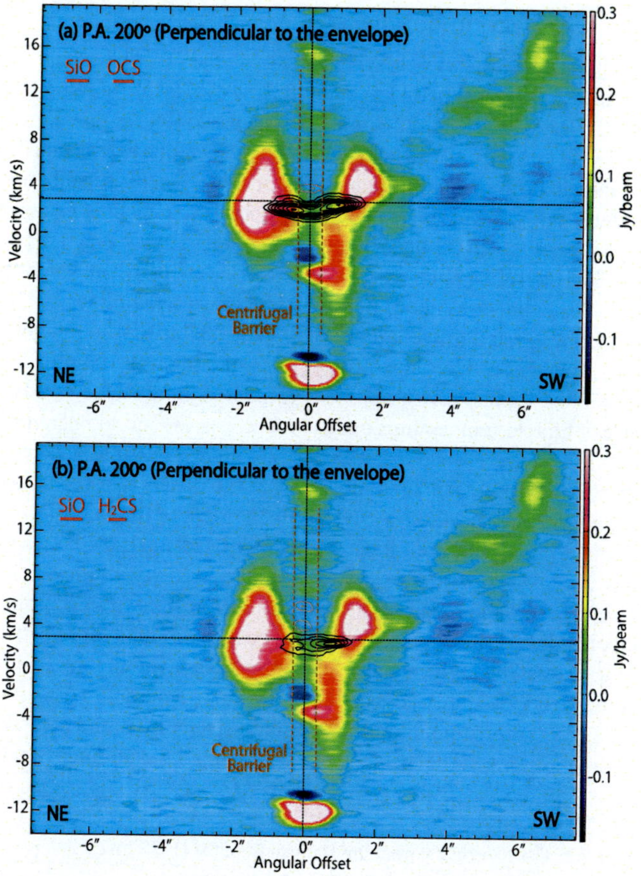

Fig. 7.18 Position-velocity diagrams of the SiO ($J = 6 - 5$; color), OCS ($J = 19 - 18$; contours in panel **a**), and H_2CS ($7_{0,7} - 6_{0,6}$; contours in panel **b**) lines. Position axes are taken to be perpendicular to the disk/envelope direction (P.A. 200°), and centered at the protostellar position. The absorption features at the velocity of -2 and -10 km s^{-1} in the color map seem to be the contamination by other unidentified molecular lines

from those reported previously. The complex structure of this binary system may cause this contradiction. The dynamical timescale of the outflow lobes is estimated to be from 10^4 to 10^5 years at a larger scale ($10^4 - 10^5$ au; [4, 12]). On the other hand, it is reported that Source A and Source B are rotating around each other with the period of $\sim 2 \times 10^4$ years [1]. Thus, the directions of the two outflow can be modulated in a complex way. Hence, to explore the relation between the outflow and the disk/envelope system, the outflow structure at a small-scale in the vicinity of the protostar should be studied.

7.7 Gas Kinetic Temperature

In Chap. 6, the gas kinetic temperature was evaluated from the intensities of H_2CS lines in IRAS 16293−2422 Source A. The gas kinetic temperature around Source A was found to rise from the infalling-rotating envelope to the centrifugal barrier once, and then to drop in the disk component inside the centrifugal barrier.

Here, the gas kinetic temperature around Source B is evaluated based on the intensities of the H_2CS lines. Figure 7.11 shows the PV diagrams of the high excitation lines of H_2CS ($7_{2,5} - 6_{2,4}$; $7_{4,4} - 6_{4,3}$, $7_{4,3} - 6_{4,2}$), on which those of the H_2CS ($7_{0,7} - 6_{0,6}$) line are superposed. The higher excitation lines of H_2CS seem to have more concentrated distributions to the protostellar position in comparison with the lowest excitation line ($7_{0,7} - 6_{0,6}$). Thus, the intensity ratio of a higher excitation line relative to the lowest one rises as approaching to the protostar, implying that the gas kinetic temperature does so. The gas kinetic temperature is evaluated by using the non-LTE code RADEX [19], as shown in Table 7.2. Temperatures for the five positions are calculated: positions at the distance of 0, 40, and 80 au from the continuum peak position along the disk/envelope direction (P.A. 110°). They are calculated based on the ratios of the integrated intensity between the three H_2CS lines ($7_{0,7} - 6_{0,6}$; $7_{2,5} - 6_{2,4}$; $7_4 - 6_4$). The velocity width for the integration is 1 km s^{-1}, where the range of the velocity shift differs from position to position. The velocity-shift range for the continuum peak position is set to be from −2 to −1 km s^{-1} to extract the high velocity-shift component avoiding the absorption effect. That for the positions at the radius of 40 au from the continuum peak position is from ±1 to 0 km s^{-1}, corresponding to the rotating velocity at the centrifugal barrier. That for the positions at the distance of 80 au from the continuum peak position is from

Table 7.2 Gas Kinetic Temperature Derived from the H_2CS Lines[a]

	Offset from the protostar[b]				
Transitions	−80 au[c]	−40 au[d]	0 au[e]	+40 au[f]	+80 au[c]
$7_{2,5} - 6_{2,4}$ /$7_{0,7} - 6_{0,6}$[g]	120 − 171	> 306	> 167	> 228	85 − 109
$7_4 - 6_4$/$7_{2,5} - 6_{2,4}$[h]	55 − 118	85 − 118	73 − 400	< 85	63 − 120

[a]In K. The gas kinetic temperatures are derived by using the RADEX code [19]. The H_2 density is assumed to be within the range from 10^7 to 3×10^9 cm^{-3}, while the column density of H_2CS to be within the range from 10^{13} to 10^{15} cm^{-2}. The quoted errors represent 1σ estimated from the statistical error. The calibration error is not taken into account, because it is expected to be almost canceled in the intensity ratios

[b]The positive and negative values represent the offset from the continuum peak position toward the southeastern and northwestern direction, respectively

[c–f]The velocity-shift range for integration is (c) from −0.5 to +0.5, (d) from −1.0 to 0.0, (e) from −2.0 to −1.0, and (f) from 0.0 to +1.0 km s^{-1}

[g]The intensity ratio between the two H_2CS lines ($7_{0,7} - 6_{0,6}$; $7_{2,5} - 6_{2,4}$) is used to derive the gas kinetic temperature

[h]The intensity ratio between the two H_2CS lines ($7_{2,5} - 6_{2,4}$; $7_{4,4} - 6_{4,3}$, $7_{4,3} - 6_{4,2}$) is used to derive the gas kinetic temperature. Note that the $7_{4,4} - 6_{4,3}$ and $7_{4,3} - 6_{4,2}$ lines have the same rest frequency

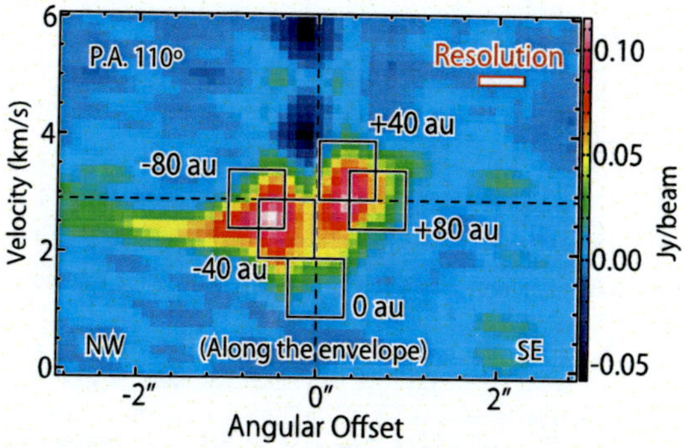

Fig. 7.19 Position-velocity diagram of the H_2CS ($7_{0,7} - 6_{0,6}$) line prepared along the disk/envelope direction (P.A. 110°). Black rectangles represent where the integrated intensities are calculated to derive the gas kinetic temperature in Sect. 7.7. Their vertical edges represent the velocity range for the integration. Their horizontal edges are centered at the positions at the distance of 0, 40, and 80 au from the continuum peak position, and their lengths correspond to the spatial resolution for the H_2CS ($7_{0,7} - 6_{0,6}$) line observation

-0.5 to $+0.5$ km s^{-1}. The above positions and velocity-ranges for the integration are depicted in the PV diagram of the H_2CS ($7_{0,7} - 6_{0,6}$) line (Fig. 7.19). It should be noted that the H_2CS ($7_{0,7} - 6_{0,6}$) line possibly be optically thick near the continuum peak position. Hence, its integrated intensity would be attenuated there, resulting in possible overestimation of the gas kinetic temperature. Although no systematic dependence on the radius from the protostar can be found, the gas kinetic temperature is generally as high as ∼100 K within ∼80 au. If the dust temperature is similarly as high as the derived gas kinetic temperature, COMs can be evaporated from the dust surfaces, which will result in the hot corino chemistry of IRAS 16293−2422 Source B. No particular enhancement of the gas kinetic temperature was found at the centrifugal barrier in Source B in contrast to the Source A case. This is probably because of the limited resolution of the observations, and thus, higher angular resolution observations are required for further investigation.

7.8 Abundance of $HCOOCH_3$ Relative to CH_3OH

The column densities of CH_3OH and $HCOOCH_3$ are derived for the five regions along the disk/envelope direction, where the gas kinetic temperature is derived with H_2CS lines. The velocity width for integration is the same as the H_2CS case for the evaluation of the gas kinetic temperature (Sect. 7.7). Hence, the derived column densities do not represent the total ones, and hence, they do not have quantitative meanings by themselves. However, the abundance ratios $HCOOCH_3/CH_3OH$ are meaningful for comparison from position to position. In IRAS 16293−2422 Source

Table 7.3 Abundance of $HCOOCH_3$ Relative to CH_3OH[a]

	Offset from the protostar[b]				
T_{rot}^{c}	−80 au[d]	−40 au[e]	0 au[f]	+40 au[g]	+80 au[d]
50 K	5.2 ± 0.5	3.2 ± 0.1	2.0 ± 0.2	5.3 ± 1.2	2.7 ± 0.3
100 K	2.5 ± 0.2	1.54 ± 0.06	0.98 ± 0.09	2.5 ± 0.6	1.3 ± 0.2
150 K	2.1 ± 0.2	1.29 ± 0.05	0.82 ± 0.08	2.1 ± 0.5	1.1 ± 0.1

[a]The quoted errors represent 3σ, where σ is derived from the statistical error. The velocity ranges from 0.9 km s^{-1} to 2.9 km s^{-1}.
[b]Position offsets toward the southeastern and northwestern direction from the continuum peak position are represented by the positive and negative values, respectively.
[c]Assumed rotational temperature.
[d–g]The velocity-shift range for the integrated intensities is (d) from −0.5 to +0.5, (e) from −1.0 to 0.0, (f) from −2.0 to −1.0, and (g) from 0.0 to +1.0 km s^{-1}

A, it was found that the abundance ratio shows a systematic increase as approaching to the protostar position (Chap. 6). This may suggest that the chemical composition is processed to enhance $HCOOCH_3$ in the vicinity of the protostar in Source A.

The derived abundance ratios in Source B are listed in Table 7.3. This analysis relies on the LTE assumption. In addition, the rotation temperatures are assumed to be 50, 100, and 150 K, based on the range of the gas kinetic temperature derived from the above H_2CS analysis (Sect. 7.7). If the CH_3OH line is assumed to be optically thin, the $HCOOCH_3$/CH_3OH abundance ratio is found to be almost equal to or higher than unity. Such a high ratio is consistent with previous reports (e.g. [2]). Optical depths of the CH_3OH and $HCOOCH_3$ lines are estimated to be 0.61 and 0.48, respectively, for the region offset from the protostellar position by −40 au, where their integrated intensities take their highest values, for the rotation temperature of 50 K. Thus, the optically thin assumption is appropriate. In contrast to the Source A case (Chap. 6), the ratio seems lower in the central part than those in the outer part. The ratios are generally lower than those of the disk component in Source A (8.7 – 8.9).

7.9 Summary of This Chapter

The OCS, CH_3OH, $HCOOCH_3$, H_2CS, and SiO lines were observed toward IRAS 16293–2422 Source B at a sub-arcsecond resolution with ALMA. Major findings are as follows:

(1) The above molecular lines show chemical differentiations similar to those found in Source A (Chap. 6); the OCS and H_2CS lines have more extended distributions than the CH_3OH and $HCOOCH_3$ lines.
(2) A marginal rotation feature was found in the extended component of the disk/envelope system of Source B traced by the H_2CS and OCS lines, in spite of its nearly face-on configuration. The observed kinematic structure is reasonably interpreted in terms of the infalling-rotating envelope model assuming the

ballistic motion. In contrast, any rotation motion was found in the CH_3OH and $HCOOCH_3$ lines.

(3) The bipolar outflow lobes near the protostar were detected in the SiO line, where the blue and red lobes are largely overlapped. This suggests a nearly pole-on geometry with a slight inclination.

(4) Absorption features were found toward the protostellar position in the observed molecular lines, except for the SiO line. They are interpreted as the inverse P-Cygni profile due to the infalling gas along the line of sight, as previously reported. The inferred infall motion is reasonably explained by assuming the free-fall motion of the gas stagnated around the centrifugal barrier.

(5) The above results are interpreted as the coexistence of the outflow and the infall motion toward the protostellar position along the line of sight. This may suggest that the current launching point of the outflow lobes responsible for the SiO emission has an offset from the protostar.

(6) The gas kinetic temperature was evaluated to be as high as 100 K by using the H_2CS lines. It is consistent with the hot corino character of this source.

(7) The abundance ratio of $HCOOCH_3$ relative to CH_3OH was evaluated around the protostar. It is higher than unity, and is slightly lower than that in Source A.

References

1. Bottinelli S, Ceccarelli C, Neri R, Williams JP, Caux E, Cazaux S, Lefloch B, Maret S, Tielens AGGM (2004) Near-arcsecond resolution observations of the hot corino of the solar-type protostar IRAS 16293–2422. ApJ 617(1):L69–L72
2. Bottinelli S, Ceccarelli C, Williams JP, Lefloch B (2007) Hot corinos in NGC 1333-IRAS4B and IRAS2A. A&A 463(2):601–610
3. Chandler CJ, Brogan CL, Shirley YL, Loinard L (2005) IRAS 16293–2422: proper motions, jet precession, the hot core, and the unambiguous detection of infall. ApJ 632(1):371–396
4. Hirano N, Mikami H, Umemoto T, Yamamoto S, Taniguchi Y(2001) SIO emission in the multilobe outflow associated with IRAS 16293–2422. ApJ 547(2):899–906
5. Jørgensen JK, Bourke TL, Nguyen Luong Q, Takakuwa S (2011) Arcsecond resolution images of the chemical structure of the low-mass protostar IRAS 16293-2422. An overview of a large molecular line survey from the submillimeter array. A&A 534:A100
6. Jørgensen JK, van der Wiel MHD, Coutens A, Lykke JM, Müller HSP, van Dishoeck EF, Calcutt H, Bjerkeli P, Bourke TL, Drozdovskaya MN, Favre C, Fayolle EC, Garrod RT, Jacobsen SK, Öberg KI, Persson MV, Wampfler SF (2016) The ALMA protostellar interferometric line survey (PILS). First results from an unbiased submillimeter wavelength line survey of the Class 0 protostellar binary IRAS 16293-2422 with ALMA. A&A 595:A117
7. Jørgensen JK, Favre C, Bisschop SE, Bourke TL, van Dishoeck EF, Schmalzl M (2012) Detection of the simplest sugar, glycolaldehyde, in a solar-type protostar with ALMA. ApJ 757(1):L4
8. Knude J, Hog E (1998) Interstellar reddening from the HIPPARCOS and TYCHO catalogues. I. Distances to nearby molecular clouds and star forming regions. A&A 338:897–904
9. Kuan Y-J, Huang H-C, Charnley SB, Hirano N, Takakuwa S, Wilner DJ, Liu S-Y, Ohashi N, Bourke TL, Qi C, Zhang Q (2004) Organic molecules in low-mass protostellar hot cores: submillimeter imaging of IRAS 16293–2422. ApJ 616(1):L27–L30
10. Lee C-F, Mundy LG, Reipurth B, Ostriker EC, Stone JM (2000) CO outflows from young stars: confronting the jet and wind models. ApJ 542(2):925–945

11. Machida MN, Hosokawa T (2013) Evolution of protostellar outflow around low-mass protostar. MNRAS 431(2):1719–1744
12. Mizuno A, Fukui Y, Iwata T, Nozawa S, Takano T (1990) A remarkable multilobe molecular outflow: rho ophiuchi east, associated with IRAS 16293–2422. ApJ 356:184
13. Müller HSP, Schlöder F, Stutzki J, Winnewisser G (2005) The cologne database for molecular spectroscopy, CDMS: a useful tool for astronomers and spectroscopists. J. Molecular Struct 742(1-3):215–227
14. Pickett HM, Poynter RL, Cohen EA, Delitsky ML, Pearson JC, Müller HSP (1998) Submillimeter, millimeter and microwave spectral line catalog. J Quant Spec Radiat Transf 60(5):883–890
15. Pineda JE, Maury AJ, Fuller GA, Testi L, García-Appadoo D, Peck AB, Villard E, Corder SA, van Kempen TA, Turner JL, Tachihara K, Dent W (2012) The first ALMA view of IRAS 16293-2422. Direct detection of infall onto source B and high-resolution kinematics of source A. A&A 544:L7
16. Adams M, Warmels R, Remijan A (2015) ALMA cycle 3 technical handbook
17. Rodríguez LF, Loinard L, D'Alessio P, Wilner DJ, Ho PT (2005) IRAS 16293-2422B: a compact, possibly isolated protoplanetary disk in a class 0 object. ApJ 621(2):L133–L136
18. Sakai N, Oya Y, Higuchi AE, Aikawa Y, Hanawa T, Ceccarelli C, Lefloch B, López-Sepulcre A, Watanabe Y, Sakai T, Hirota T, Caux E, Vastel C, Kahane C, Yamamoto S (2017) Vertical structure of the transition zone from infalling rotating envelope to disc in the Class 0 protostar, IRAS 04368+2557. MNRAS 467(1):L76–L80
19. van der Tak FFS, Black JH, Schöier FL, Jansen DJ, van Dishoeck EF (2007) A computer program for fast non-LTE analysis of interstellar line spectra. With diagnostic plots to interpret observed line intensity ratios. A&A 468(2):627–635
20. Zapata LA, Loinard L, Rodríguez LF, Hernández-Hernández V, Takahashi S, Trejo A, Parise B (2013) ALMA 690 GHz observations of IRAS 16293–2422B: infall in a highly optically thick disk. ApJ 764(1):L14

Chapter 8
L483

8.1 Introduction

So far, it has been demonstrated that the infalling-rotating envelope and its centrifugal barrier will exist in low-mass protostellar sources regardless of their chemical characteristics. However, the chemical change occurring across the centrifugal barrier was confirmed with different molecular species depending on the chemical composition of the source. In the previous chapters, sources with peculiar chemical compositions were dealt with; warm carbon-chain chemistry (WCCC) and hot corino sources. They show the exclusive chemical compositions with each other. Thus more general understandings of the chemical change requires such chemical diagnostics in other sources with different chemical characteristics. In this chapter, a hybrid chemical characteristics of L483 is confirmed. The kinematic structure of the gas in L483 is also investigated.

It is of fundamental importance to characterize the chemical composition in the vicinity of the protostar, because it will define chemical heritage from interstellar matters to protoplanetary disks. However, observational studies of chemical characterization have been performed at a 100 au scale only for a several sources so far, including four hot corinos (IRAS 16293−2422 Source A, Source B, IRAS 2A, B335; Chaps. 6, 7; [12, 18]) and three WCCC sources (L1527, IRAS 15398−3359, TMC−1A; Chaps. 4, 5; [27, 28, 30]). Therefore, observations for other protostellar sources, especially for ones with intermediate chemical compositions, are awaited. In this chapter, a well-studied Class 0 protostellar source L483 is focused on in this context.

L483 is a dark cloud located in the Aquila Rift (d = 200 pc [15, 25]). It harbors the Class 0 protostar IRAS 18148−0440 [4, 7], and its bolometric luminosity is reported to be 13 $L_\odot$ [32]. The systemic velocity of 5.5 km s^{-1} is employed in this study based on previous single-dish observations [10]. In this source, the C_4H

This chapter has been published in Oya et al., 2017, ApJ, 837, 174 and Oya et al., 2018, ApJ, 863, 72. © AAS. Reproduced with permission.

Y. Oya, *A Few Tens au Scale Physical and Chemical Structures Around Young Low-Mass Protostars*, Springer Theses, https://doi.org/10.1007/978-981-19-1708-0_8

abundance is relatively high, and it is regarded as a possible candidate for the WCCC source [10, 11, 29, 31]. The carbon-chain-molecule rich nature of L483 has been further supported by the detection of the carbon-chain radical HCCO [1].

The outflow of L483 has extensively been studied [7, 9, 14, 17, 23, 34, 35]. It is known to be extended along the east-west axis; the eastern and western lobes are red-shifted and blue-shifted, respectively. The position angle of the outflow axis is reported to be 95° by [23] based on the HCO^+ ($J = 1 - 0$) observation and 105° by [4] based on the shocked H_2 emission reported by [7]. Its inclination angle is reported to be ~50° (0° for a pole-on configuration) [7].

The gas on the northern and southern sides of L483 are blue-shifted and red-shifted, respectively, according to the CS ($J = 2 - 1, 7 - 6$) and HCN ($J = 4 - 3$) observations [14, 35]. This velocity gradient is perpendicular to the outflow axis, and thus, it suggests a rotating motion of the envelope gas. The position angle of the pseudo-disk of L483 is reported to be 36° based on *Spitzer* 4.5 μm observation [4].

In the previous studies described above, the disk/envelope system of L483 was not well spatially resolved. The chemical composition of the gas in the closest vicinity of the protostar is still highly unknown. In this study, the physical and chemical structures around the protostar are investigated at a 100 au scale with ALMA.

8.2 Observation

L483 was observed with ALMA during its Cycle 2 operation on 12 June 2014. Band 6 receiver was employed to observe the spectral lines of CCH, CS, SO, HNCO, t-HCOOH, CH_3CHO, NH_2CHO, $HCOOCH_3$, $(CH_3)_2O$, and SiO in the frequency range from 244 to 246 GHz. Their line properties are summarized in Table 8.1. Thirty-four antennas were used during the observation with the baseline length ranging from 18.5 to 644 m. The field center of the observation was $(\alpha_{2000}, \delta_{2000}) = (18^h17^m29^s.910, -04°39'39''.60)$, and the primary beam size (FWHM) is 23''.03. The total on-source time was 25.85 min. A typical system temperature was from 60 to 100 K. A backend correlator was tuned to a resolution of 61.030 kHz, corresponding to the velocity resolution of 0.073 km s^{-1} at 250 GHz, and a bandwidth of each chunk of 58.5892 MHz. The bandpass calibrations were carried out with J1733−1304, as well the phase calibrations for every 7 min. Titan was observed to derive the absolute flux density scale. The data calibration was performed in the antenna-based manner, where the uncertainties are expected to be less than 9%.

The continuum and line images were obtained with the CLEAN algorithm. The Brigg's weighting with the robustness parameter of 0.5 was employed. Self-calibration was carried out for the phase and amplitude by using the continuum data and was applied to the continuum and line data. A continuum image was prepared by averaging line-free channels, and the line maps were obtained after subtracting the continuum component directly from the visibilities. Synthesized-beam sizes for the spectral lines are summarized in Table 8.1. An rms noise level for the continuum is 0.13 mJy $beam^{-1}$. Those for spectral line images are obtained in nearby line-free

Table 8.1 Parameters of the Observed Lines[a]

Molecule	Transition	Frequency (GHz)	$E_u k_B^{-1}$ (K)	$S\mu^2$[b] (D^2)	A_{ij} (s^{-1})	Beam Size
1.2 mm continuum						0″.46 × 0″.42 (P.A. 11.°76)
CS	$J = 5 - 4$	244.9355565	35.3	19	2.98×10^{-4}	0″.51 × 0″.46 (P.A. −177.°24)
NH_2CHO[c]	$12_{0,12} - 11_{0,11}$	247.390719	78.1	156	1.10×10^{-3}	0″.56 × 0″.49 (P.A. 16.°44)
$HCOOCH_3$[c]	$20_{5,16,0} - 19_{5,15,0}$	249.0474280	141.6	50	1.46×10^{-3}	0″.52 × 0″.45 (P.A. −177.°18)
SiO[c]	$J = 6 - 5$	260.5180090	43.8	58	9.12×10^{-4}	0″.46 × 0″.42 (P.A. −177.°78)
CH_3CHO[c]	$14_{1,14} - 13_{1,13}$; A	260.5440195	96.4	82	6.25×10^{-4}	0″.46 × 0″.42 (P.A. −177.°81)
SO	$J_N = 6_7 - 5_6$	261.8437210	47.6	16	2.28×10^{-4}	0″.46 × 0″.42 (P.A. 3.°09)
CCH[d]	$N = 3 - 2$, $J = 7/2 - 5/2$, $F = 4 - 3$	262.0042600	25.1	2.3	5.32×10^{-5}	0″.98 × 0″.92 (P.A. −78.°06)
	$F = 3 - 2$	262.0064820	25.1	1.7	5.12×10^{-5}	0″.98 × 0″.92 (P.A. −78.°06)
t-HCOOH[c]	$12_{0,12} - 11_{0,11}$	262.103481	82.8	24	2.03×10^{-4}	0″.46 × 0″.42 (P.A. 2.°98)
$(CH_3)_2O$[c]	$13_{5,8,1} - 13_{4,9,1}$	262.393513	118.0	148	7.18×10^{-5}	0″.46 × 0″.41 (P.A. 1.°69)
HNCO[c]	$12_{0,12} - 11_{0,11}$	263.7486250	82.3	30	2.56×10^{-4}	0″.48 × 0″.41 (P.A. 3.°87)

[a] Taken from CDMS [20] and JPL [24]
[b] Nuclear spin degeneracy is not included
[c] Four channel bind. The spectral profile of the t-HCOOH line (Fig. 8.10) is obtained with binding 16 channels
[d] An outer taper of 1″ is applied

channels with the channel width of 61.030 kHz; 8.2, 7.6, 6.1, 6.5, 4.4, 5.8, and 3.5 mJy beam^{-1} for the CCH, CS, SO, HNCO, NH_2CHO, $HCOOCH_3$, and SiO emissions, respectively.

8.3 Distribution

The 1.2 mm dust continuum map is shown in Fig. 8.1. Its distribution is fitted by a two-dimensional Gaussian profile. Its peak position is derived to be: $(\alpha_{2000}, \delta_{2000}) = (18^h17^m29^s.947, -04°39'39''.55)$. Its size deconvolved by the beam is derived to be $0''.23 \times 0''.16$ (P.A. 158°), and thus, the continuum image is scarcely resolved in this observation. Its total flux is 28 mJy.

Various molecular lines are detected in L483: CCH, CS, SO, HNCO, NH_2CHO, $HCOOCH_3$, and SiO. This source is regarded as a WCCC candidate source [10, 11, 29, 31], as mentioned in Sect. 8.1. Hence, detections of complex organic molecules (COMs), such as NH_2CHO and $HCOOCH_3$, which are characteristic to hot corinos (e.g. [3, 31]), are notable. The integrated intensity maps of CCH, CS, SO, HNCO, NH_2CHO, $HCOOCH_3$, and SiO are shown in Figs. 8.2, 8.3 and 8.4.

The CCH ($N = 3 - 2, J = 7/2 - 5/2, F = 4 - 3$ and $3 - 2$) emission is extended over a 10″ scale (Fig. 8.2a). Thus, the outer taper of 1″ is applied to improve a signal-to-noise ratio of the image. The WCCC nature of L483 is confirmed by the

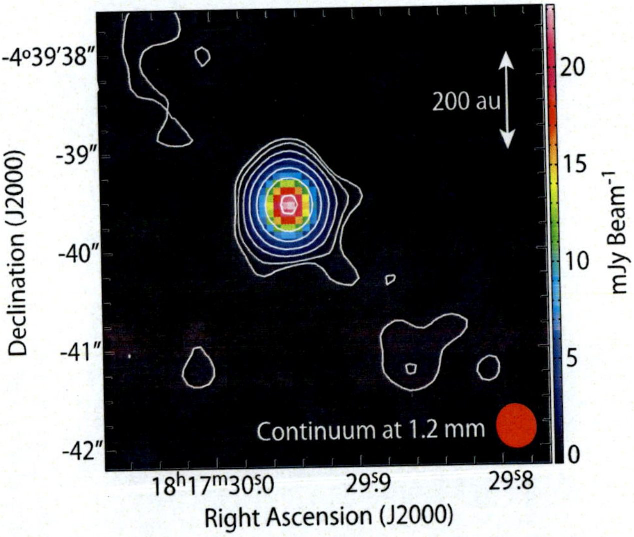

Fig. 8.1 1.2 mm dust continuum map. Contour levels are 3, 5, 10, 20, 40, 80, and 160σ, where the rms noise level is 0.13 mJy beam^{-1}. The beam ($0''.46 \times 0''.42$; P.A. 11.°76) is denoted in the bottom right corner

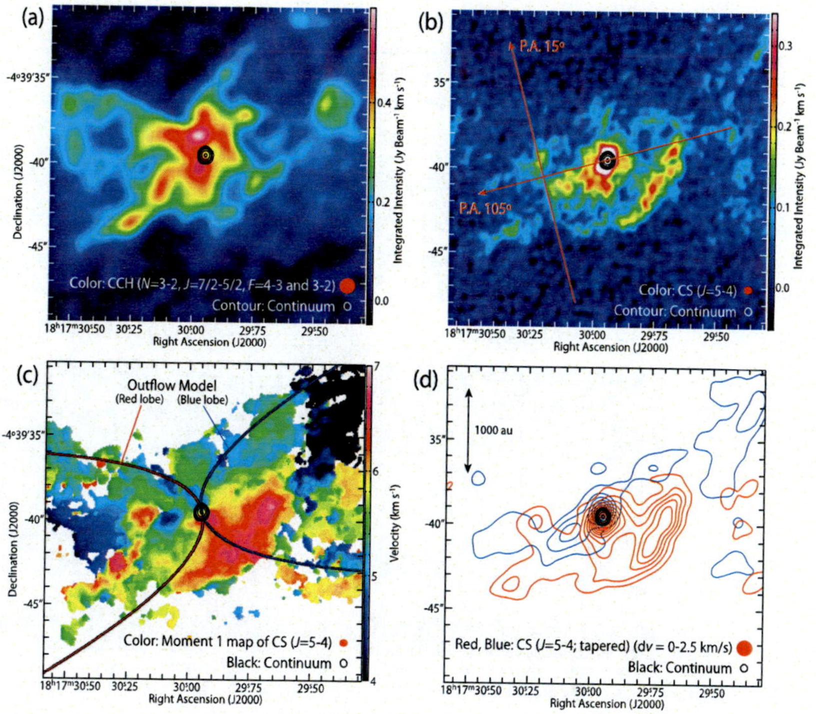

Fig. 8.2 Integrated intensity maps of the CCH ($N = 3 - 2$, $J = 7/2 - 5/2$, $F = 4 - 3$ and $3 - 2$; **a**) and CS ($J = 5 - 4$; **b**, **d**) lines, and the velocity map of the CS line (**c**). Black contours represent the 1.2 mm dust continuum map. Contour levels are 10, 20, 40, 80, and 160σ, where the rms noise level is 0.13 mJy beam^{-1}. The velocity-shift range for the integration is ± 8 km s^{-1} for panels (**a**) and (**b**) with respect to the systemic velocity (5.5 km s^{-1} [10]). The CCH emission has the contribution from the two hyperfine components (see Table 8.1). Panel (**d**) shows the low velocity-shift components of the CS line with an outer taper of 1″, where the velocity-shift range for the integration is from 0 to +2.5 km s^{-1} (red contours) and from −2.5 to 0 km s^{-1} (blue contours). Contour levels for the CS emissions are every 5σ from 3σ, where the rms noise level is 20 mJy beam^{-1} km s^{-1}. The red arrow with a P.A. of 105° in panel (**b**) represents the outflow axis. The red arrow with a P.A. of 15° represents the position axis along which the PV diagrams in Fig. 8.8b, c are prepared. It is perpendicular to the outflow axis and centered at the position with an offset of 4″ to the southeast from the continuum peak along the outflow axis (P.A. 105°)

existence of the carbon-chain molecule CCH around the protostar at a few 100s au scale. The CCH emission has an intensity dip near the protostellar position with a radius of ∼0.″5. This feature is similar to that found in the WCCC sources L1527 and IRAS 15398−3359 (Chap. 5; [28, 30]), which would originate from the gas-phase destruction and/or depletion onto dust grains of CCH. The hole of the intensity distribution seems to have a slight offset from the continuum peak position to the western side, implying asymmetry in the gas distribution in the vicinity of the protostar. This asymmetry may be related to inhomogeneities of the initial gas distribution. A part of the CCH emission seems to trace the outflow cavity walls extending along

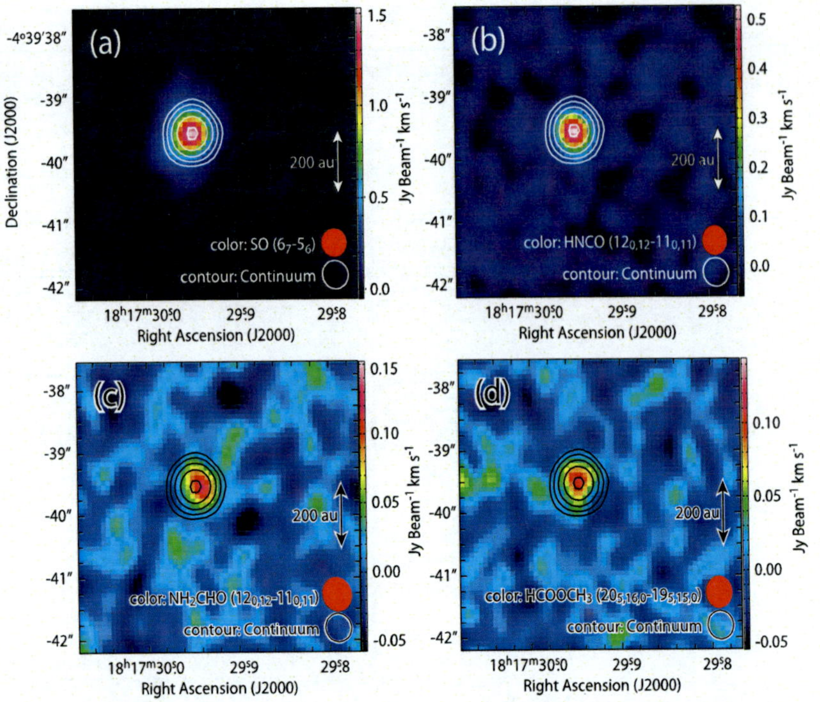

Fig. 8.3 Integrated intensity maps of the SO ($J_N = 6_7 - 5_6$; **a**), HNCO ($12_{0,12} - 11_{0,11}$; **b**), NH_2CHO ($12_{0,12} - 11_{0,11}$; **c**), and $HCOOCH_3$ ($20_{5,16,0} - 19_{5,15,0}$; **d**) emissions. Contours represent the 1.2 mm continuum map, where the contour levels are the same as those in Fig. 8.2

the southeast-northwest direction, as well as the envelope component. The direction of the outflow axis looks consistent with the previous reports (P.A. 95°–105°; e.g. [4, 23, 34]).

As well as CCH, the CS ($J = 5 - 4$) emission traces the component extended over a 10″ (Fig. 8.2b). In addition, it shows a compact component concentrated to the continuum peak position. The deconvolved size of this compact component is derived to be 1.″26 × 0.″88 by the two-dimensional Gaussian fitting; it is slightly more extended than the 1.2 mm dust continuum. This slightly extended component will be discussed in Sect. 8.4. Meanwhile, the distribution of the SO ($J_N = 6_7 - 5_6$), HNCO ($12_{0,12} - 11_{0,11}$), NH_2CHO ($12_{0,12} - 11_{0,11}$), and $HCOOCH_3$ ($20_{5,16,0} - 19_{5,15,0}$) emissions are highly concentrated to the continuum peak position (Fig. 8.3). Their deconvolved sizes are derived by the two-dimensional Gaussian fittings; 0.″56 × 0.″39, and 0.″26 × 0.″16 for SO and HNCO, respectively. The HNCO distribution is almost point-like. Similarly, the distributions of NH_2CHO and $HCOOCH_3$ are point-like with the angular resolution of ∼0.″5 (∼100 au). In addition to these COM emissions, the t-HCOOH emission is tentatively detected concentrated near the protostellar position.

The SiO ($J = 6 - 5$) emission has a different distribution from all the above molecular emissions. As shown in Fig. 8.4, the SiO distribution is slightly extended

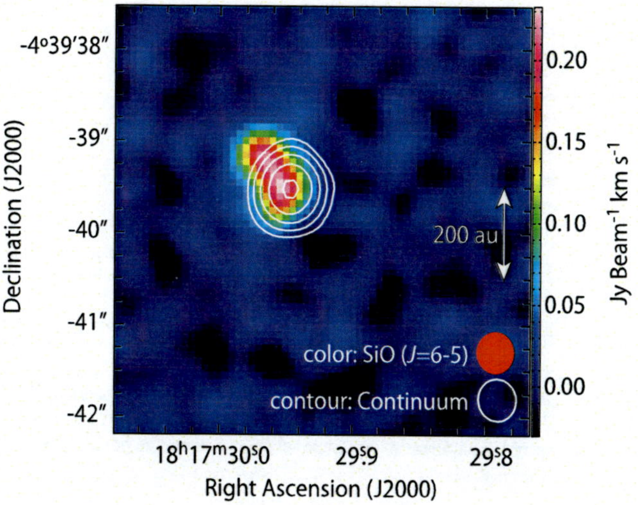

Fig. 8.4 Integrated intensity map of the SiO ($J = 6 - 5$) emission. White contours represent the 1.2 mm continuum map, where the contour levels are the same as those in Fig. 8.2

toward the northeastern direction from the continuum peak position. The extension is significant, considering the beam size of $\sim0''.5$ ($\sim$100 au). The deconvolved size of the SiO distribution is derived to be $0''.70 \times 0''.43$ (P.A. 35°) by the two-dimensional Gaussian fitting.

8.4 Kinematic Structure

8.4.1 Geometrical Configuration

This section focuses on the kinematic structure of the gas concentrated around the protostar. The velocity (moment 1) maps of the SO and HNCO emissions (Fig. 8.5) show velocity gradients perpendicular to the outflow axis (P.A. 95°–105°; e.g. [4, 23, 34]), which strongly suggests rotating motion of the gas; the gas on the northern side of the continuum peak position is blue-shifted, while that on the southern side is red-shifted. The direction of this velocity gradient is qualitatively consistent with that at a 20″ scale reported by [35] based on the CS ($J = 7 - 6$) and HCN ($J = 4 - 3$) observations. In this study, the position axis (P.A.) of 105° is employed as the direction of the outflow axis according to the report by [4], and the mid-plane of the disk/envelope system is assumed to extend along the P.A. of 15°. The eastern and western lobes of the bipolar outflow are reported to be red- and blue-shifted, respectively (e.g. [7, 9, 14, 17, 23, 34, 35]), and thus, the northwestern side of the disk/envelope system is thought to face the observer. This geometrical configuration is schematically illustrated in Fig. 8.6. The following subsections describe the kinematic structures traced by the observed molecular line emissions, except for CCH because of the heavy blending of its hyperfine components and poor S/N ratio.

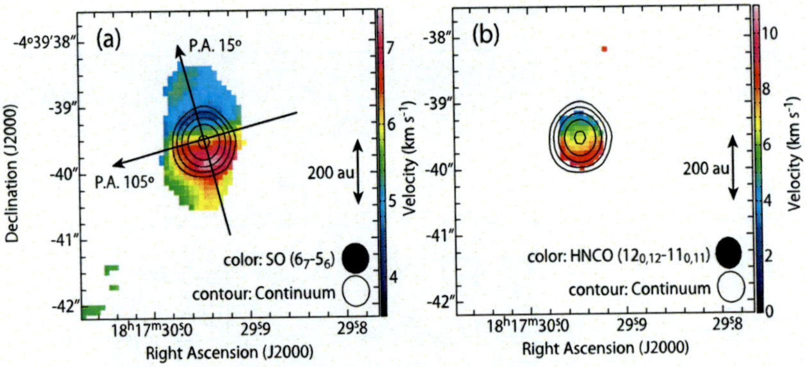

Fig. 8.5 Velocity (moment 1) maps of the SO ($J_N = 6_7 - 5_6$; **a**) and HNCO ($12_{0,12} - 11_{0,11}$; **b**) emissions. Black contours represent the 1.2 continuum map, where the contour levels are the same as those in Fig. 8.2. The black arrows in panel (**a**) represent the position axes of the PV diagrams in Figs. 8.7, 8.9, 8.11–8.14

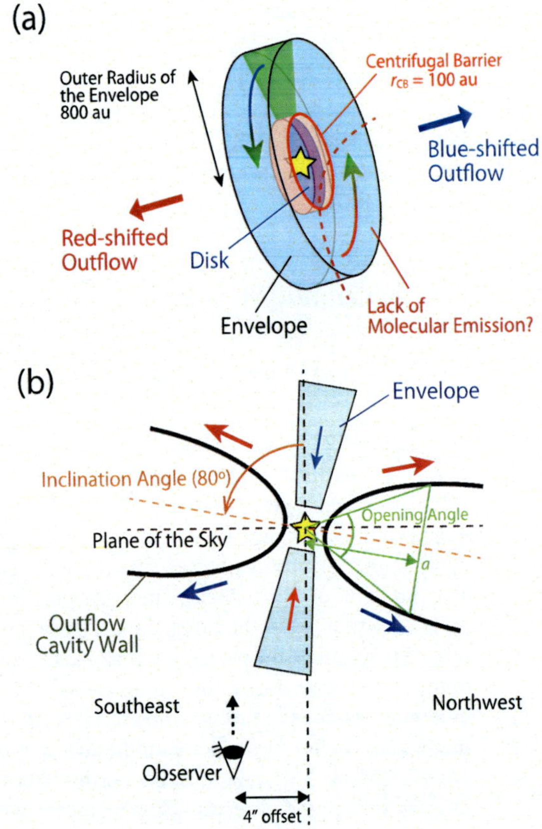

Fig. 8.6 Schematic illustrations of the disk/envelope system in L483. In panel (**a**), the line of sight is perpendicular to the paper. The mid-plane of the disk/envelope is extended along the northeast-southwest direction with a P.A. of 15°. Its western side faces the observer. Line emissions in the western side of the protostar are missing in the observation (Sect. 8.3). As shown in panel (**b**), the outflow blows nearly parallel to the plane of the sky so that both the red- and blue-shifted components are seen in each lobe. The PV diagrams in Fig. 8.8b, c are prepared for the positions offset from the protostellar position by 4″ as indicated by the 'observer'

8.4.2 CS

The position-velocity (PV) diagrams of the CS ($J = 5 - 4$) line are shown in Fig. 8.7. Figure 8.7a, c are prepared along the mid-plane of the disk/envelope system ('the disk/envelope direction'; P.A. 15°). Spin-up features to the protostar are seen along the disk/envelope direction on an 8″ scale; the gas is red- and blue-shifted on the southwestern and northeastern side, respectively. Meanwhile, the counter velocity component is weakly seen, although the blue-shifted component on the southwestern

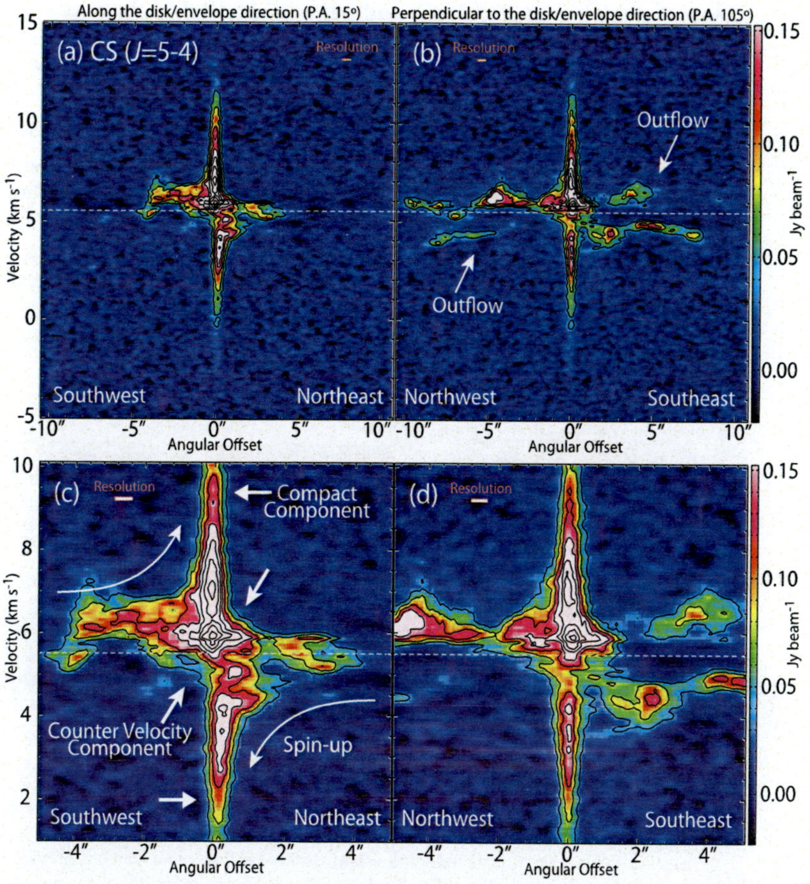

Fig. 8.7 Position-velocity diagrams of the CS ($J = 5 - 4$) line. The position axes are indicated by the arrows in Fig. 8.5a, which are taken along the disk/envelope direction (P.A. 15°; **a**, **c**) and perpendicular to it (P.A. 105°; **b**, **d**). Panels (**c**) and (**d**) are the blow-ups of the central parts of panels (**a**) and (**b**). Contour levels are every 5σ, where the rms noise level is 7.6 mJy beam^{-1}. White dashed lines represent the systemic velocity (5.5 km s^{-1}). The rectangle in the top-right or top-left corner of each panel represents the spatial and velocity resolutions

side is marginal. This feature is specific to the infalling-rotating motion, as demonstrated in the previous chapters (e.g. Chaps. 3 and 4). Hence, CS likely exists in the infalling-rotating envelope. In addition, high velocity-shift components are detected concentrated to the protostar, whose maximum velocity-shift is as high as about 6 km s^{-1}.

An extended component over 15″ is seen along the line perpendicular to the disk/envelope direction (P.A. 105°; Fig. 8.7b). Although this component looks complicated, it likely traces a part of the outflow cavity. The PV diagrams of the outflow are shown in Fig. 8.8. Figure 8.8a is prepared along the outflow axis, which is a part of Fig. 8.7b. Figure 8.8b shows the PV diagram of the southeastern outflow lobe prepared along the line across the outflow axis indicated by the arrow in Fig. 8.2b (P.A.15°). In Fig. 8.8a, it is confirmed that the northwestern and southeastern lobes are blue- and red-shifted, respectively. Figure 8.8a also shows that the velocity of the outflow component accelerates as the distance from the protostar.

The elliptic feature of the PV diagram characteristic to the outflow cavity wall is clearly seen in Fig. 8.8b. Similar features are confirmed in the two hyperfine components of CCH (Fig. 8.8c). This elliptic feature has both the red- and blue-shifted velocities, where its central velocity is slightly red-shifted. This kinematic structure suggests the configuration of the outflow illustrated in Fig. 8.6b. It is quite similar to those observed in the nearly edge-on outflow system of IRAS 15398−3359 (Chap. 5), where both the red-shifted and blue-shifted components can be seen in each outflow lobe. Therefore, it is most likely that the outflow of L483 blows nearly perpendicular

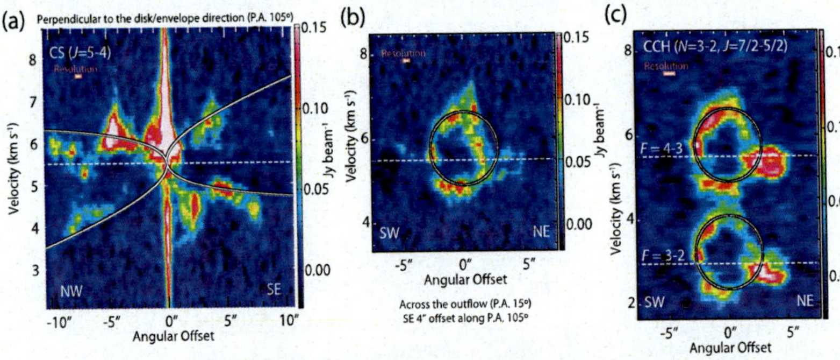

Fig. 8.8 Position-velocity diagrams of the CS ($J = 5-4$; **a**, **b**) and CCH ($N = 3-2$, $J = 7/2-5/2$, $F = 4-3$ and $3-2$; **c**) lines. The position axes are indicated by the arrows in Fig. 8.2b. Panel (**a**) is a part of Fig. 8.7b, which is prepared along the outflow axis (P.A. 105°). Panels (**b**) and (**c**) show the kinematic structure of the southeastern outflow lobe, which is prepared along the line across the outflow axis (P.A. 15°). Their position axes are centered at the distance of 4″ from the protostellar position toward the southeastern direction. White lines represent the results of the parabolic outflow model. The parameters for the model are as follows; $i = 80°$, $C = 2.5 \times 10^{-3}$ au^{-1}, and $v_0 = 1.5 \times 10^{-3}$ km s^{-1}. In panel (**c**), the two hyperfine components of CCH are detected with the velocity separation of 2.54 km s^{-1}, and the elliptic feature appears twice. Thus, the outflow model is prepared for each hyperfine component

to the line of sight at least in the vicinity of the protostar in contrast to the previous reports (e.g. [7]). Hence, the disk/envelope system likely has a nearly edge-on geometry. The more detailed structure of the outflow is described in Sect. 8.6.1.

Assuming the configuration illustrated in Fig. 8.6a for the disk/envelope system, the northwestern side of the disk/envelope system will face the observer. If there is an infall motion in the envelope component, the southeastern and northwestern sides of the protostar would be red-shifted and blue-shifted, respectively. However, this feature is not clearly recognized in the PV diagram along the P.A. of 105° (Fig. 8.7d). This can be attributed to overwhelming contributions from the outflow component and the missing of the blue-shifted envelope component. The velocity gradient due to the infall motion will be verified for the low velocity-shift component ($|v_{\rm shift}| < 2$ km s^{-1}) with the aid of the kinematic model in Sect. 8.5. On the other hand, the high velocity-shift components ($|v_{\rm shift}| > 2$ km s^{-1}) do not show any velocity gradient along the P.A. of 105° around the protostellar position.

8.4.3 SO and HNCO

Figure 8.9 shows the PV diagrams of the SO ($J_N = 6_7 - 5_6$) and HNCO ($12_{0,12} - 11_{0,11}$) lines. The SO emission shows a slight velocity gradient along the disk/envelope direction (Fig. 8.9a). This feature is scarcely recognized in the HNCO emission (Fig. 8.9c). The velocity gradient in the SO emission is consistent with that found in the CS emission, and therefore, it likely originates from the rotating motion in the disk/envelope system. Meanwhile, no definitive velocity gradient is seen along the line perpendicular to the disk/envelope direction either in the SO and HNCO emissions (Fig. 8.9b, d). Hence, these emissions do not reveal a significant infall motion in the vicinity of the protostar.

All the SO, HNCO, and CS emission have high velocity-shift components near the protostellar position in their PV diagrams (Figs. 8.7–8.9). These components are confirmed in the line profiles of these molecular lines prepared in a circular region with a diameter of $0''.5$ centered at the continuum peak (Fig. 8.10). The SO emission has a line profile similar to the CS emission, except for the self-absorption feature in the CS emission. Their broad line-widths are recognized as the high velocity-shift components concentrated to the protostar, which are found in the PV diagrams.

The HNCO emission has a component with a velocity-shift larger than 5 km s^{-1}; the red-shifted component is brighter than the blue-shifted component. This feature can be confirmed in both its PV diagrams (Figs. 8.9c, d) and line profile (Fig. 8.10). It implies that the HNCO distribution is asymmetric in the vicinity of the protostar. Alternatively, the blue-shifted emission from the back side of the protostar may be attenuated by dust in the closest vicinity of the protostar. This effect is discussed for the similar feature found in the Class I low-mass protostellar source Elias 29 [21].

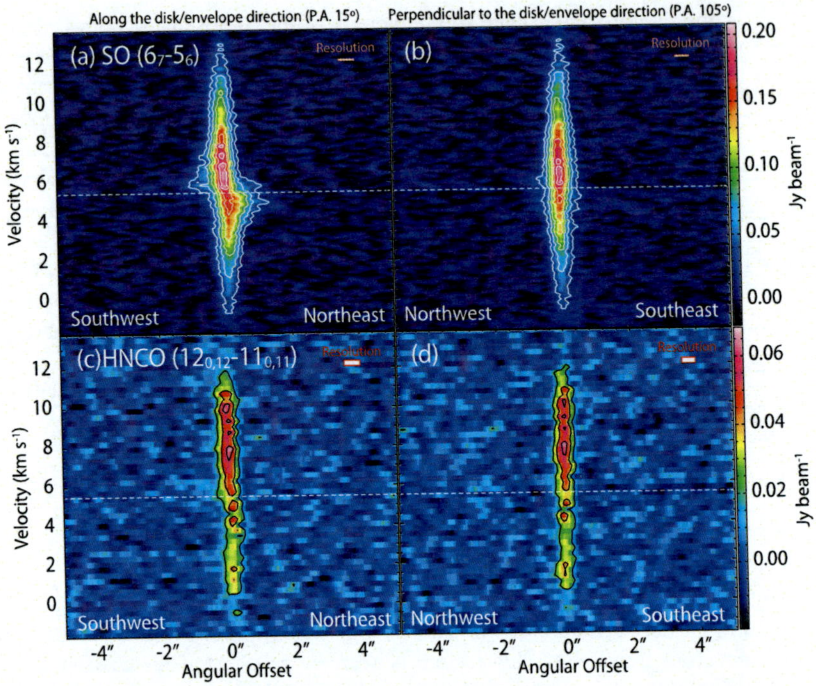

Fig. 8.9 Position-velocity diagrams of the SO ($J_N = 6_7 - 5_6$; **a**, **b**) and the HNCO ($12_{0,12} - 11_{0,11}$; **c**, **d**) lines. The position axes are as the same as those in Fig. 8.7c, d. Contour levels are every 5 and 3σ, where the rms noise levels are 6.1 and 6.5 mJy beam^{-1}, for SO and HNCO, respectively

8.4.4 *NH_2CHO and $HCOOCH_3$*

It is notable that the COM emissions, NH_2CHO and $HCOOCH_3$, are successfully detected in L483. Their distributions are concentrated to the protostar, as shown in their integrated intensity maps (Figs. 8.3c, d). Their spectral line profiles toward the protostellar position are dominated by a red-shifted component (Fig. 8.10), as well as the profile of the HNCO line. The consistency between the profiles of these molecular lines supports their secure detections rather than a possibility that they correspond to other molecular lines. As described in Sect. 8.4.3, the line profiles biased to the red-shifted components imply asymmetric distributions of these molecules in the vicinity of the protostar and/or the attenuation of the blue-shifted components by dust.

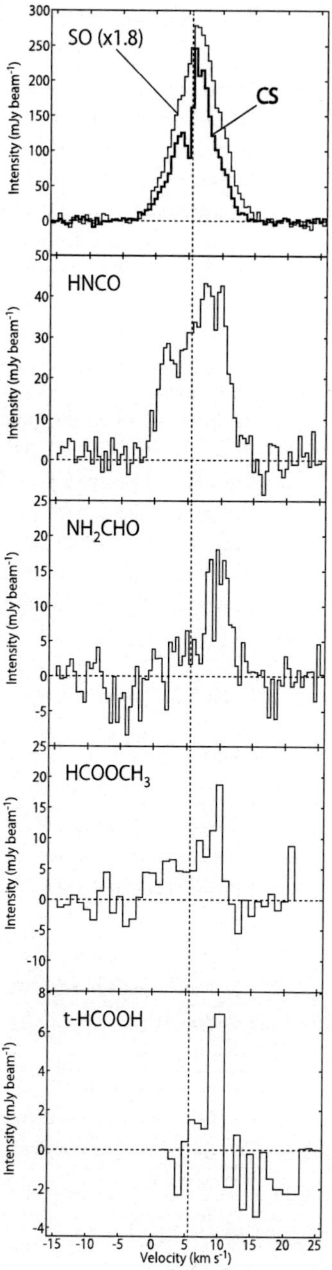

Fig. 8.10 Spectral line profiles of the CS ($J = 5 - 4$), SO ($J_N = 6_7 - 5_6$), HNCO ($12_{0,12} - 11_{0,11}$), NH_2CHO ($12_{0,12} - 11_{0,11}$), $HCOOCH_3$ ($20_{5,16,0} - 19_{5,15,0}$), and t-HCOOH ($12_{0,12} - 11_{0,11}$) lines toward the protostellar position. The line intensities are averaged in a circular region with a diameter of 0.″5 centered at the continuum peak position. The spectra are smoothed to improve the signal-to-noise ratio so that the velocity resolution is changed to be 0.5 km s^{-1} for the CS, SO, HNCO, and NH_2CHO lines, and 1 km s^{-1} for the $HCOOCH_3$ and t-HCOOH lines

8.5 Analysis with the Models for the Disk/Envelope System

The molecular distributions described above show chemical differentiation in L483. To understand it in terms of the physical structure around the protostar, the observed kinematic structure of the disk/envelope system is analyzed in this section. In previous chapters, the kinematic structures of the infalling-rotating envelopes are successfully explained by a simple ballistic model in several sources. Hence, the same model is applied to the CS emission observed in L483. However, it is found that an infalling-rotating envelope model cannot reproduce all the observed velocity structure. Two physical components need to be considered as described below; the infalling-rotating envelope and the centrally concentrated component.

The infalling-rotating envelope model described in Chap. 3 is employed; a flat envelope with a constant thickness (30 au) is assumed. The inclination angle (i) of 80° is employed (Table 8.2), which is obtained based on the analysis of the outflow structure (Sect. 8.6). It is unfortunate that the key parameters for the model are loosely constrained based on these observations because of the contamination of the overwhelming centrally-concentrated component. Hence, models with various parameters are calculated to find the reasonable set of the parameters by eye. An example of the infalling-rotating envelope model is shown in Fig. 8.11a, b, which reproduces the observed PV diagrams as much as possible except for the centrally-

Table 8.2 Physical Parameters of L483

Protostar	
Age[a]	$< 4.2 \times 10^3$ years
	(with an uncertainty of a factor of three)
Bolometric luminosity ($L_{\rm bol}$)[b]	13 $L_\odot$
Bolometric temperature ($T_{\rm bol}$)[a]	< 56 K
Protostellar mass (M)[c]	0.15 $M_\odot$
Disk/Envelope	
Radius of the centrifugal barrier ($r_{\rm CB}$)[c]	100 au
Specific angular momentum ($j_{\rm IRE}$)[c]	$7.9^{+4}_{-3} \times 10^{-4}$ km s^{-1} pc
Outflow	
Position angle[d]	105°
Inclination angle (i)[e,f]	80°
C for the parabolic outflow model[e]	2.5×10^{-3} au^{-1}
v_0 for the parabolic outflow model[e]	1.5×10^{-3} km s^{-1}
Dynamical timescale ($t_{\rm dyn}$)[e]	$(3 \pm 1) \times 10^3$ years

[a] Taken from [7]
[b] Taken from [32]
[c] See Sect. 8.5
[d] Taken from [4]
[e] See Sect. 8.6.2
[f] 0° for a pole-on configuration

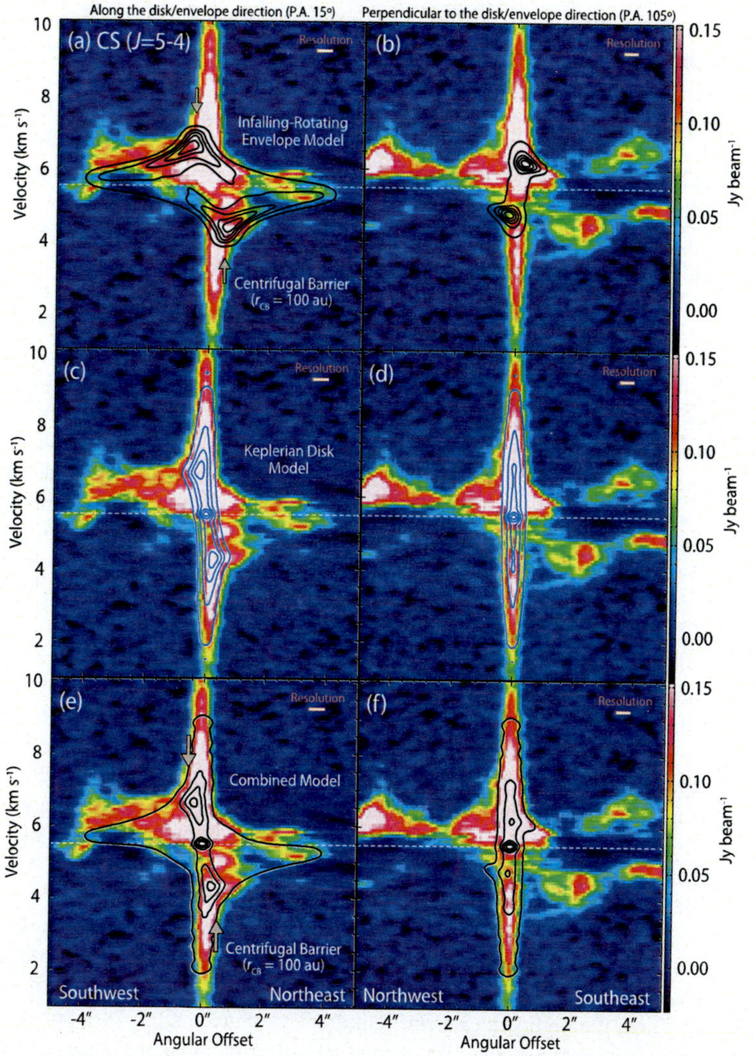

Fig. 8.11 Position-velocity diagrams of the CS ($J = 5-4$) line and the fiducial models. Color maps are the same as those in Fig. 8.7c, d. Black contours in panels (**a**) and (**b**) represent the results of the infalling-rotating envelope model. The parameters of the model are as follows: M is 0.15 $M_{\odot}$, $r_{\rm CB}$ is 100 au, and i is 80°. Blue contours in panels (**c**) and (**d**) represent the results of the Keplerian disk model. The above M and i values are employed, and the emitting region is assumed to be inside the centrifugal barrier. Black contours in panels (**e**) and (**f**) represent the results combining the infalling-rotating envelope model and the Keplerian disk model. The intrinsic line width of 0.2 km s^{-1} is employed for all the models, and the model images are convolved with the Gaussian beam for the CS line in the observation (Table 8.1). Contour levels are every 20% from 5% (**a–d**) or 3% (**e, f**) of the peak intensity in each panel

concentrated high velocity-shift components. The parameters of this model are as follows: the protostellar mass (M) is 0.15 $M_\odot$ and the radius of the centrifugal barrier (r_{CB}) is 100 au. Then, the specific angular momentum (j_{IRE}) of the gas is calculated to be $7.9^{+4}_{-3} \times 10^{-4}$ km s^{-1} pc. These values are summarized in Table 8.2. With the aid of the model result in Fig. 8.11b, the infall motion of the gas is marginally recognized. Its blue-shifted part is missing in the observation probably due to the asymmetric gas distribution described above (Fig. 8.6). A possible effect of the attenuation of blue-shifted components by dust is expected to be modester for positions slightly shifted from the protostar than toward the protostellar position.

The dependency of the infalling-rotating envelope model result on the protostellar mass (M) and the radius of the centrifugal barrier (r_{CB}) are examined by calculating the models with various sets of these two parameters. The results are shown in Figs. 8.12 and 8.13. The models with M of 0.05 $M_\odot$ or 0.5 $M_\odot$ do not reproduce the observed PV diagrams with any values of r_{CB}. In the model without any rotating motion, i.e. r_{CB} is 0 au, no velocity gradient appears along the disk/envelope direction (Fig. 8.12). This contradicts with the observed feature. Meanwhile, the velocity shift of the counter-velocity component due to the infall motion is underestimated with r_{CB} of 300 au. Based on the above results, a reasonable agreement between the observation and the model is obtained for the range of M from 0.1 $M_\odot$ to 0.2 $M_\odot$ and r_{CB} from 30 au to 200 au, except for the centrally-concentrated high velocity-shift components in the observation. Hence, the infalling-rotating envelope model with M of 0.15 $M_\odot$ and r_{CB} of 100 au is employed as the fiducial one, which is shown in Fig. 8.11. To obtain more stringent constraints on the parameters, further detailed analysis with a higher angular-resolution observation is required.

The infalling-rotating envelope model cannot reproduce both the high velocity-shift components concentrated to the protostar and the extended low velocity-shift components simultaneously; if the parameters are set to explain the high velocity-shift components, the infall motion in the model is calculated to be much larger than the observed emission. This justifies the two-component model consisting of the infalling-rotating envelope and the centrally concentrated component. Although the observed red-shifted emission shows some excess on the southwestern side to the protostar (Fig. 8.11a), the fiducial model roughly reproduce the infalling-rotating envelope component in the PV diagrams.

Then, the high velocity-shift components concentrated to the protostar are analyzed in a separated way. The Keplerian disk component is the most likely candidate for them, although the rotation signature is not spatially resolved. The results of the Keplerian disk model are superposed on the PV diagrams of the CS and SO lines in Figs. 8.11 and 8.14, repsectively. M of 0.15 $M_\odot$ and i of 80° are employed in this model as the fiducial values based on the analyses for the envelope structure described above and for the outflow structure in Sect. 8.6. The observed emission has the maximum velocity-shift as high as 6 km s^{-1}, which corresponds to the emitting region of the radius as small as 4 au in the Keplerian disk model. Moreover, the results combining the infalling-rotating envelope model and the Keplerian disk model are shown in Fig. 8.11e, f. The results of the Keplerian disk model with various parameter values are also shown in Figs. 8.12 and 8.13 for reference.

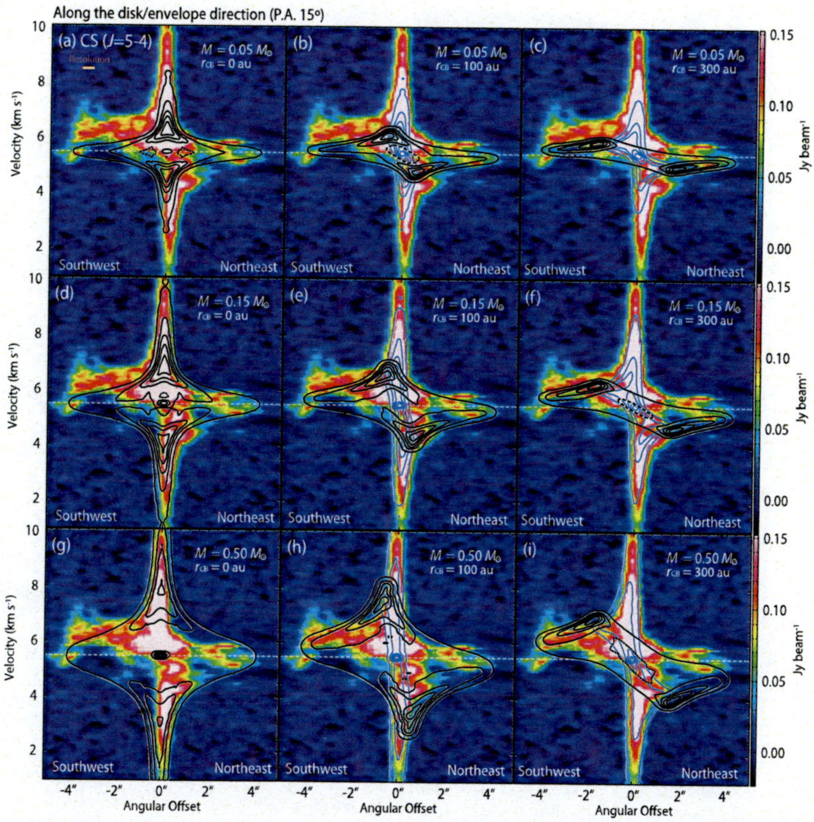

Fig. 8.12 Comparison of the position-velocity diagrams of the observed CS ($J = 5 - 4$) line and the infalling-rotating envelope model. Color maps are the same as that in Fig. 8.7c, where the position axis is along the disk/envelope direction. Black contours represent the results of the infalling-rotating envelope model with the following parameters: M is 0.05, 0.15, and 0.5 $M_{\odot}$, $r_{\rm CB}$ is 0, 100, and 300 au, and i is fixed to be 80°. Blue contours in panels represent the results of the Keplerian disk model with the same parameters as those for the infalling-rotating envelope model in each panel. In the Keplerian disk model, the emission is simply assumed to come from the inside of the centrifugal barrier. Then, the Keplerian disk model is not prepared for panels (**a**), (**d**), and (**g**), where $r_{\rm CB}$ is 0 au. Contour levels are every 20% from 5% of the peak intensity of each model. Dashed contours around the central position in panels (**a**), (**b**), (**d**), (**f**), (**h**), and (**i**) represent the dip in the modeled intensity

In the infalling-rotating envelope model and the Keplerian disk model, the emissivity is assumed to be proportional to $r^{-1.5}$ for simplicity, where r denotes the distance from the protostar. Then, the effects of the optical depth, excitation, and the temperature gradient are not considered, as described in Chap. 3. The optically thin condition can roughly be justified for the CS emission in L483 as discussed below. The observed intensity of the CS line is less than 0.2 Jy beam^{-1} at the distance of 100 au ($r_{\rm CB}$) or further from the protostar, which corresponds to the brightness tem-

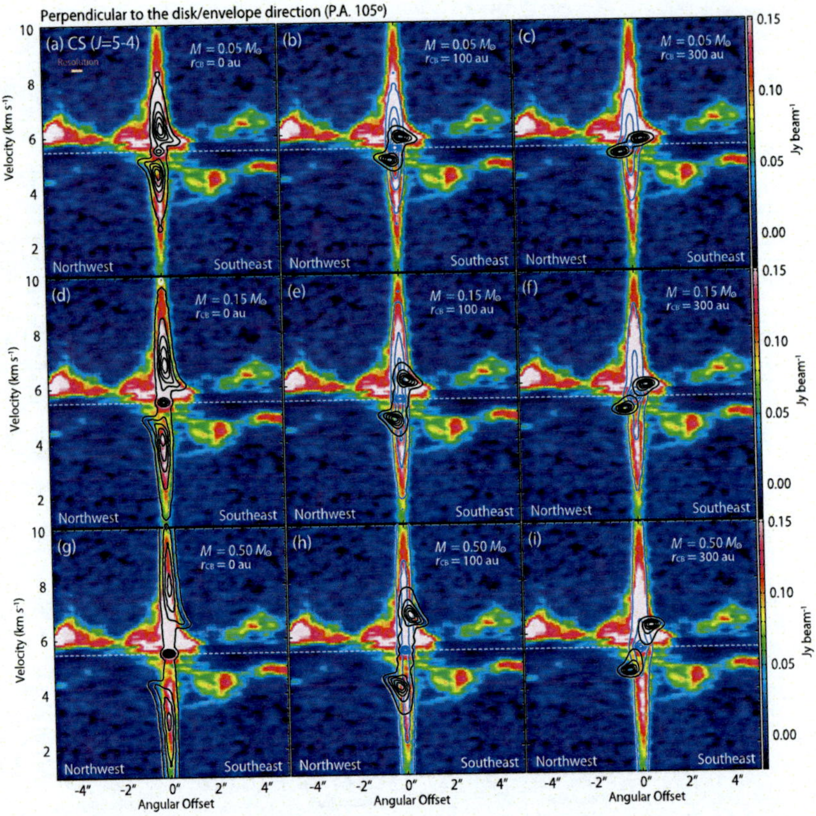

Fig. 8.13 Comparison of the position-velocity diagrams of the observed CS ($J = 5 - 4$) line and the infalling-rotating envelope model. Color maps are the same as that in Fig. 8.7d, where the position axis is perpendicular to the disk/envelope direction. Black and blue contours represent the results of the infalling-rotating envelope model and the Keplerian disk model. Their details are described in the caption of Fig. 8.12

perature less than 17 K. Meanwhile, the desorption temperature of CS is evaluated to be 35 K (see Appendix A; [37]) by using its binding energy of 1900 K (UMIST Database for Astrochemistry [19]; http://udfa.ajmarkwick.net/index.php), and thus, the gas kinetic temperature is expected to be similar to or higher than this value. Therefore, the observed brightness temperature of the CS emission is expected to be lower than the gas kinetic temperature by a factor of 2 or more. That is, the CS emission can be assumed to be optically thin in the envelope component outside the centrifugal barrier. The peak intensity of the CS line is about 0.35 Jy beam^{-1} inside the centrifugal barrier, corresponding to the brightness temperature of 30 K. The gas temperature (T) is roughly estimated to be 75 K at the distance (r) of 100 au from the protostar with the luminosity (L) of $13 L_\odot$ [32], by using the following equation:

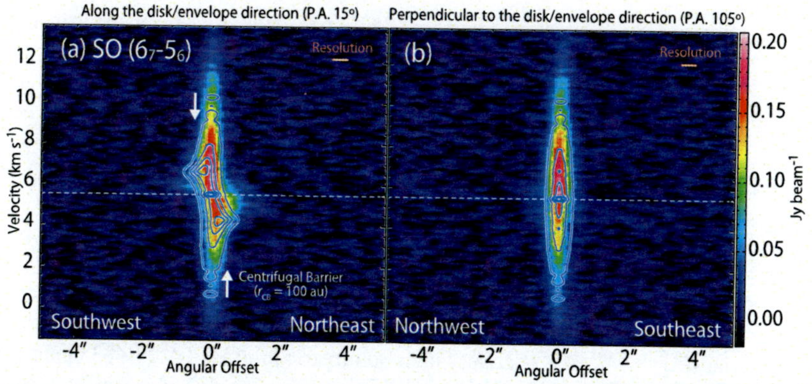

Fig. 8.14 Position-velocity diagrams of the SO ($J_N = 6_7 - 5_6$) line and the Keplerian disk model. Color maps are the same as those in Fig. 8.9a, b. The model is the same as that in Fig. 8.11c, d, other than the emission is convolved with the Gaussian beam for the SO line in the observation (Table 8.1). Contour levels are every 20% from 3% of the peak intensity in each panel

$\frac{L}{4\pi r^2} = \sigma T^4$, where σ denotes the Stefan-Boltzmann constant. Thus, the peak CS intensity is expected to be lower than the gas kinetic temperature. Therefore, the assumption of the optically thin condition for the CS emission seems reasonable. Nevertheless, the excitation effect and the temperature gradient may affect the intensity distribution in reality. These effects may have a larger contribution at positions nearer to the protostar, where the gas has a higher velocity-shift. To avoid seriously being affected by these effects, the above analyses based on the models are performed by mainly focusing on the velocity structures rather than the intensity profiles.

The kinematic structure around the protostar traced by the CS emission in L483 is similar to that traced by the H_2CS line in IRAS 16293−2422 Source A (Chap. 6), which is recognized as a combination of the infalling-rotating envelope component and the possible Keplerian disk component inside the centrifugal barrier. Hence, the CS distribution in L483 is found to be different from those in L1527 and TMC−1A (Chap. 4; [27]), where the CS emission comes only from the infalling-rotating envelope. The distribution of CS in L483 seems to resemble to that in IRAS 16293−2422 Source A [22]; in IRAS 16293−2422 Source A, the $C^{34}S$ ($J = 2 - 1$) emission was detected by ALMA in both the circummultiple structure with the infalling-rotating motion and the circumstellar disk with the Keplerian rotation, although the latter component was not detected in the $C^{34}S$ ($J = 7 - 6$) emission with the Submillimeter Array possibly due to the limitation of the angular resolution and the sensitivity [6]. The difference of the behavior of the CS lines would originate from the difference of the bolometric luminosity of the above sources; L483 (13 $L_\odot$ [32]) and IRAS 16293−2422 Source A (22 $L_\odot$ [5]) has higher luminosities than L1527 (1.7 $L_\odot$ [8]) and TMC−1A (2.5 $L_\odot$ [8]). The higher bolometric luminosity would cause the higher temperature of the disk component inside the centrifugal barrier, which prevents the CS depletion in this region. As described above, the desorption temperature of CS is evaluated to be about 35 K (Appendix A). This value is just above the temperature of the disk mid-plane just inside the centrifugal barrier in L1527 (30 K [28]). On the contrary, CS is possibly prevented from freezing out in L483 if the gas and dust temperatures inside the centrifugal barrier is higher than the desorption temperature of CS due to the high bolometric luminosity of L483.

A rotation feature is clearly revealed in the PV diagram of the SO emission (Fig. 8.9a). A similar feature was reported for L1527 [30], where the SO emission mainly highlights the centrifugal barrier. With respect to the SO line observed in L1527, that in L483 shows much brighter emission with the high velocity-shift concentrated toward the protostar. Thus, the rotating motion traced by the SO emission in L483 is expected to mainly come from the disk component inside the centrifugal barrier (Fig. 8.14). The relatively high bolometric luminosity of L483 could contribute to prevent SO from depletion onto dust grains in the disk component, as mentioned above.

The disk radius can be estimated by assuming that the SO emission appears inside the centrifugal barrier. The distribution of the SO emission has the beam-deconvolved size (FWHM) of $0\farcs5$ (100 au) along the disk/envelope direction (P.A. 15°), and thus, the radius of the centrifugal barrier is estimated to be about 50 au. Since the size of the distribution is underestimated due to the strong emission concentrated to the protostar, it can be regarded as the lower limit. Meanwhile, the 5σ contour in the PV diagram of the SO line is extended over a $2\farcs4$ (480 au) area along the disk/envelope direction (Fig. 8.9a), whose radius is obtained to be about $1''$ (200 au) by deconvolving from the beam size. Considering that the SO emission may trace the infalling-rotating envelope component just outside the centrifugal barrier [26], this radius is regarded as the upper limit for the radius of the centrifugal barrier. The above estimations will give rough estimates for the radius of the centrifugal barrier in L483, although it is difficult to be derived directly from the SO emission due to the contamination by the disk component. These values are consistent with the estimate based on the analysis of the PV diagrams of the CS emission.

8.6 Outflow Structure

8.6.1 Outflow Cavity Wall Traced by CS

The disk/envelope system of L483 is assumed to have a nearly edge-on configuration with the inclination angle (i) of 80° (0° for a face-on configuration) in the analysis of the infalling-rotating envelope described above. This value is derived from the kinematic structure of the outflow traced by the CS emission as described in this section.

The PV diagram of the CS emission (Fig. 8.8a) shows that the outflow component is accelerated as leaving from the protostar along the outflow axis. An elliptic feature is evident across the outflow axis in Fig. 8.8b. Since an extended component is often resolved out in interferometric observations, the elliptic feature likely corresponds to the outflow cavity wall. These features are quite similar to those found in a low-mass Class 0 source IRAS 15398−3359 (Chap. 5), where the kinematic structure of the outflow cavity wall was well explained by a parabolic outflow model [16]. Thus, the outflow structure in L483 is analyzed with this outflow model, whose details are described in Chap. 3.

8.6.2 Comparison with the Outflow Model

The results of the parabolic outflow model is shown by white lines in Fig. 8.8. This fiducial model reasonably explains the observed kinematic structure. In this model, the outflow cavity wall has a parabolic feature, and the velocity of the gas on the cavity wall is proportional to the distance from the protostar. The parameters for the fiducial model are as follows; i is 80°, C is 2.5×10^{-3} au^{-1}, and v_0 is 1.5×10^{-3} km s^{-1}. They are summarized in Table 8.2. This model is also superposed on the velocity map of the CS emission (Fig. 8.2c).

The observed kinematic structure cannot be reproduced by the outflow model with the inclination angle less than 75° or larger than 90°. Thus, the outflow is confirmed to blow almost parallel to the plane of the sky based on its kinematic structure. This is contrast to the previous report that L483 has the inclination angle of 50° [7], which was evaluated from the asymmetric brightness of the two outflow lobes in the near-infrared and submillimeter observations. This discrepancy in the inclination angle would come from the large uncertainty of the previous estimation. Thus, the inclination angle of 80° is employed as the fiducial value in this chapter.

Figures 8.15 and 8.16 show the PV diagrams of the CS ($J = 5-4$) emission. They are prepared along the lines perpendicular to the outflow axis. The origin of each position axis is taken on the outflow axis with an offset from 0″ to 10″ from the protostellar position along the southeast or northwest direction in Figs. 8.15 and 8.16, respectively. Although the diagrams are heavily contaminated by the emission from the disk/envelope system in the panels with the offsets from 0″ to 1″ from the protostellar position, the elliptic feature characteristic to the outflow cavity wall is confirmed in the panels with larger offsets as well as that with an offset of 4″ (Fig. 8.8b). The radial size of the elliptic feature seems to increase as leaving from the protostellar position. This indicates the expansion of the outflow cavity wall. The velocity centroid of the elliptic feature is slightly red- and blue-shifted in the southeastern and northwestern outflow lobe, respectively. This is consistent with the configuration of L483 illustrated in Fig. 8.6. The PV diagrams of the two outflow lobes with the same distance from the protostellar position are not always similar to each other; for instance, the radial size of the elliptic feature at the distance of 8″ is more compact in the southeastern lobe (Fig. 8.15) than in the northwestern lobe (Fig. 8.16). This asymmetry of the outflow structure would be caused by an asymmetry in the ambient environment.

The result of the fiducial outflow model is superposed on the observed PV diagrams in white lines in Figs. 8.17 and 8.18. The outflow model reasonably explains

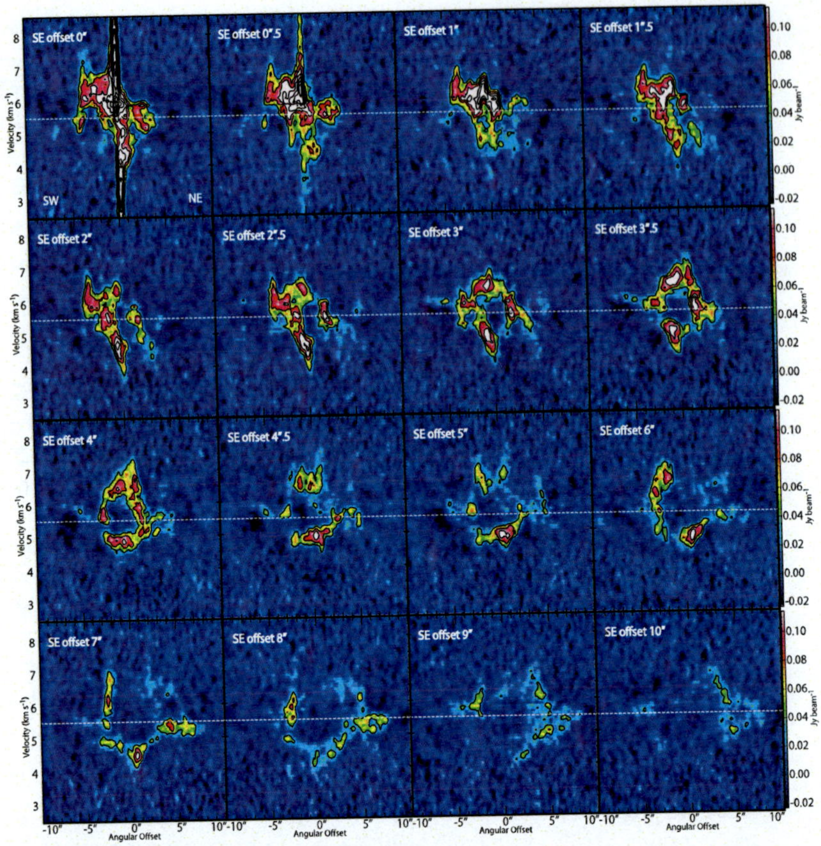

Fig. 8.15 Position-velocity diagrams of the CS ($J = 5 - 4$) emission across the southeastern outflow lobe. The position axes are prepared along the lines with the P.A. of 15°, which are perpendicular to the outflow axis (P.A. 105°). Their origins are on the outflow axis with an offset of the distance from 0″ to 10″ from the protostellar postion toward the southeastern direction. The panel labeled as 'SE offset 4″' is the same as the color map in Fig. 8.8b. The contour levels are every 5σ, where the rms noise level is 7.6 mJy beam^{-1}

the observed kinematic structure. In the southeastern outflow lobe, there is some excess of red-shifted velocity component in the panels with an offset less than 4″5. This possibly comes from a local shock on the outflow cavity wall. Such a feature is also detected in the outflow of IRAS 15398−3359 (Chap. 5).

The dynamical time scale ($t_{\rm dyn}$) of the outflow of L483 is evaluated to be $(3 \pm 1) \times 10^3$ years by using the fiducial value of v_0 (1.5×10^{-3} km s^{-1}). [9] previously reported model calculations with $t_{\rm dyn}$ of $(2 - 6) \times 10^3$ years for the CO ($J = 4 - 3, 2 - 1$) observations, and [38] reported it to be $(4.4 - 6.2) \times 10^3$ years based on the CO ($J = 6 - 5, 3 - 2$) observations. Meanwhile, [7] evaluated 13×10^3 years based on the CS ($J = 3 - 2$) observations by assuming the inclination angle of 50°.

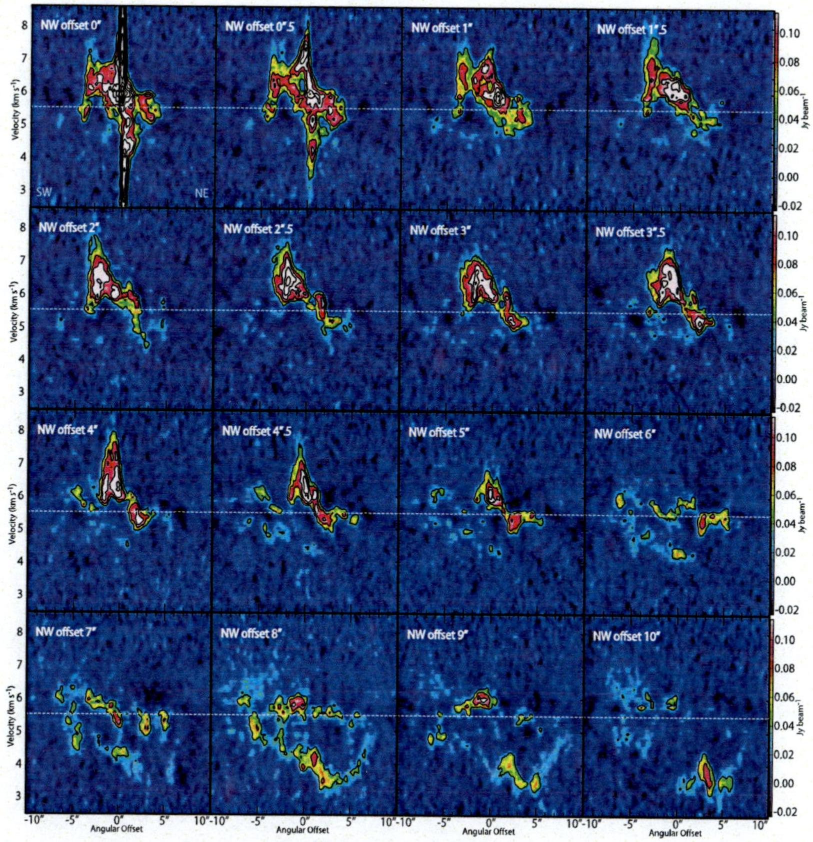

Fig. 8.16 Same as Fig. 8.15 for the northwestern outflow lobe

It is recalculated to be 2×10^3 years for the inclination angle of 80° derived from the model analysis described above. Thus, these estimations for t_{dyn} are almost consistent with one another, and a few 10^3s years is the most plausible value. It is notable that the t_{dyn} derived based on the kinematic structure observed on a 10^3 au scale in this study is consistent with the previous reports based on the observations on larger scales (e.g. 10^4 au [38]).

8.6.3 Rotation Motion in the Outflow

Outflow launching is thought to be a potential mechanism to extract the angular momentum of the envelope gas, for instance at its centrifugal barrier. If this is the case, outflows are expected to have a rotating motion. There is indeed a hint that the outflow structure of L483 has a velocity gradient perpendicular to the outflow axis

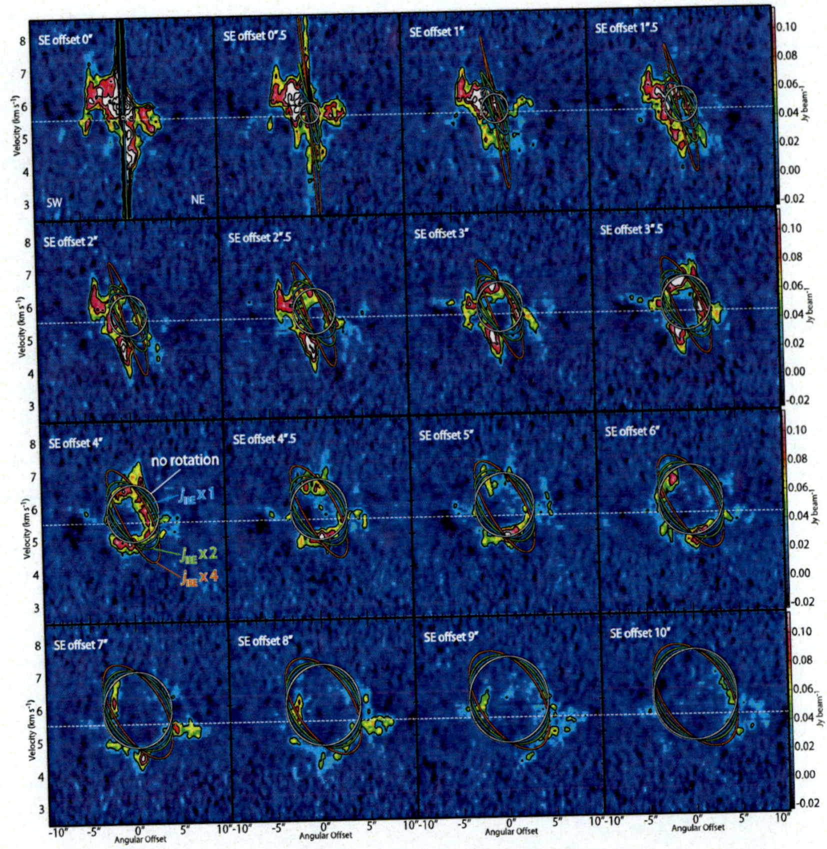

Fig. 8.17 Position-velocity diagrams of the CS ($J = 5 - 4$) emission across the southeastern outflow lobe. The color maps and black contours are the same as those in Fig. 8.15. Elliptic lines in each panel represent the results of the outflow model, where the parameters are as follows; $i = 80°$, $C = 2.5 \times 10^{-3}$ au^{-1}, and $v_0 = 1.5 \times 10^{-3}$ km s^{-1}. The specific angular momentum of the outflow model is 0, $j_{\rm IRE}\times 1$, $j_{\rm IRE}\times 2$, and $j_{\rm IRE}\times 4$ for the white, blue, green, and red lines, respectively, where the specific angular momentum of the infalling-rotating envelope ($j_{\rm IRE}$) is 7.9×10^{-4} km s^{-1} pc

(Fig. 8.2c, d). Thus, a rotating motion on the outflow cavity wall is examined in this section with the aid of the outflow model.

The blue, green, and red lines in Figs. 8.17 and 8.18 represent the results of the outflow model with some rotation motion. In these models, the rotation velocity on the outflow cavity wall is simply calculated by assuming the angular momentum conservation. The specific angular momentum of the outflowing gas ($j_{\rm out}$) is set to be 1 time (blue), 2 times (green), and 4 times (red) that of the infalling-rotating envelope ($j_{\rm IRE}$; Table 8.2). The difference between the four outflow models are more evident in panels with a smaller offset from the protostellar position. The model results with

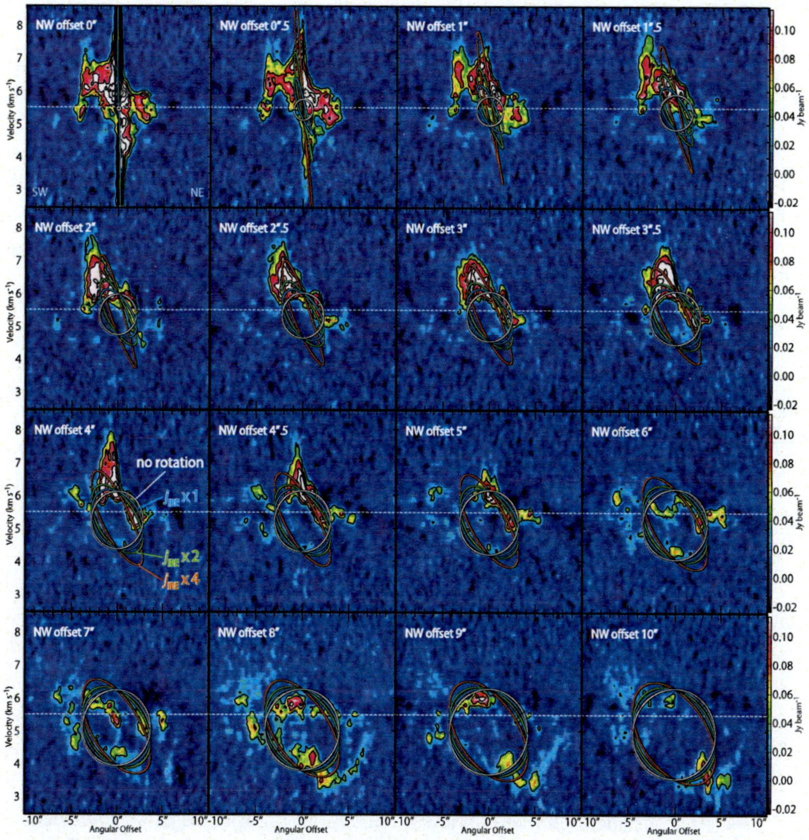

Fig. 8.18 Same as Fig. 8.17 for the northwestern outflow lobe

rotation motion show a slant distortion in the elliptic features in the PV diagrams, which is more significant with a larger j_{out}. This is because the rotation motion is more prominent for a smaller radial size under the angular momentum conservation. The models with j_{out} of 1 and 2 times j_{IRE} would better reproduce the observations than the model without the rotation motion in the panels with an offset from 1$''$ to 4$''$ from the protostellar position. The rotation motion tends to be overestimated in the model with j_{out} of 4 times j_{IRE}, and thus, this value can be recognized as the upper limit for j_{out}. Hence, the specific angular momentum of the gas on the outflow cavity wall is likely comparable to or slightly larger than that of the infalling-rotating envelope. This implies that the outflow may extract the angular momentum of the gas from the infalling-rotating envelope in L483 (see Chap. 10). Evaluation of the gas mass of these structures are required to assess the contribution of the outflow in the redistribution of the angular momentum more precisely.

8.6.4 SiO Emission

The SiO ($J = 6 - 5$) emission shows a distribution extended toward the northeastern direction from the protostellar position, as mentioned in Sect. 8.3 (Fig. 8.4). This elongation has the position angle of about 35°. The velocity channel maps of the SiO emission is shown in Fig. 8.19. While the red-shifted component weakly appears at the protostellar position, the blue-shifted component ($v_{LSR} < 3.8$ km s^{-1}) is detected at an offset of about 0.″5 (100 au). The position of the SiO blue-shifted emission is close to the expected position of the centrifugal barrier (r_{CB} of 100 au) or inside it. At this position, the velocity shift in the SiO emission is larger than that in the CS emission. Thus, the SiO emission likely comes from a possible shock caused by the outflow rather than a part of the infalling-rotating envelope. The existence of such a shocked gas near the centrifugal barrier is puzzling. It might be related to the launching mechanism of the outflow, although the association of the shocked gas with the centrifugal barrier has to be explored at a higher angular-resolution.

8.7 Chemical Composition

L483 is proposed to have the WCCC character on the basis of the single-dish observations [10, 11, 29, 31]. It is indeed confirmed by the detection of CCH on a 1000 au scale in Fig. 8.2. Meanwhile, emission of COMs characteristic to hot corino chemistry are also detected; NH_2CHO and $HCOOCH_3$ are firmly detected, and t-HCOOH is tentatively detected. This is the first spatially-resolved detection of COMs in a WCCC source, and thus L483 can be recognized as a '*hybrid*' source. Although such a chemical characteristic source can be interpreted as an intermediate source in previous studies [31], the present observation confirms its definitive existence at a high angular resolution.

The column densities of NH_2CHO and $HCOOCH_3$ are evaluated as summarized in Table 8.3; they are the beam averaged values at the protostellar position. The local thermodynamic equilibrium (LTE) condition is assumed with the temperature of 70, 100, and 130 K, which are typical values in hot corinos (e.g. Chap. 6). The column densities of NH_2CHO and $HCOOCH_3$ are obtained to be $(1.5 \pm 0.7) \times 10^{14}$ cm^{-2} and $(7 \pm 4) \times 10^{15}$ cm^{-2}, respectively, at 100 K. The results for NH_2CHO and $HCOOCH_3$ change by 30% and 14%, respectively, with the change in the assumed temperature by ±30 K. The column densities of t-HCOOH, CH_3CHO, and $(CH_3)_2O$ are also obtained as summarized in Table 8.3, where the values for CH_3CHO and $(CH_3)_2O$ are their upper limits.

The beam averaged column density of H_2 is derived to be 6.5×10^{23} cm^{-2} from the dust continuum emission by by using the following relation [36]:

$$N(\mathrm{H_2}) = \frac{2 \ln 2 \cdot c^2}{\pi h \kappa_\nu m} \times \frac{F(\nu)}{\nu^3 \theta_{\mathrm{major}} \theta_{\mathrm{minor}}} \times \left(\exp\left(\frac{h\nu}{kT}\right) - 1 \right). \quad (8.1)$$

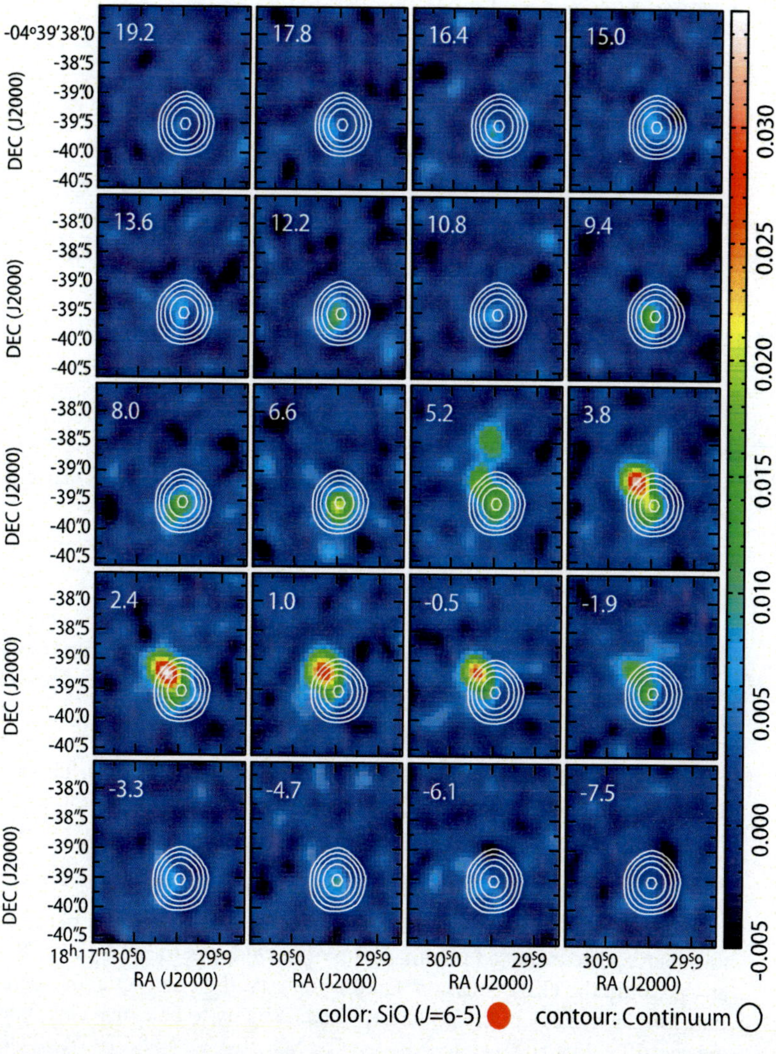

Fig. 8.19 Velocity channel maps of the SiO ($J = 6 - 5$) emission. Contours represent the 1.2 mm continuum map, where the contour levels are the same as those in Fig. 8.2. Each map represents the averaged intensity with the velocity width of 1.4 km s^{-1}. The averaged velocity for each panel is denoted in its top left corner in a unit of km s^{-1}

κ_ν the mass absorption coefficient with respect to the gas mass, m the averaged mass of a particle in the gas (3.83×10^{-24} g), ν the frequency, $F(\nu)$ the peak flux, $\theta_{\rm major}$ and $\theta_{\rm minor}$ the major and minor beam size, respectively, c the speed of light, h the Planck's constant, k the Boltzmann constant, and T the dust temperature. The dust temperature is assumed to be 100 K. κ_ν is evaluated to be 0.008 cm^2 g^{-1} at 1.2 mm with $\beta =$

Table 8.3 Column densities and fractional abundances of the molecules observed toward the protostar position[a]

	Column density (10^{14} cm^{-2})			Fractional abundance relative to H_2 (10^{-10})		
Dust temperature	70 K	100 K	130 K	70 K	100 K	130 K
CS	5.1 ± 0.2	6.2 ± 0.2	7.5 ± 0.2	5.3 ± 0.2	9.6 ± 0.3	15.2 ± 0.5
HNCO	9.6 ± 1.1	11.5 ± 1.3	14.1 ± 1.6	10.0 ± 1.1	17.7 ± 2.0	28.6 ± 3.2
NH_2CHO	1.2 ± 0.6	1.5 ± 0.7	1.9 ± 0.9	1.3 ± 0.6	2.4 ± 1.1	3.9 ± 1.9
$HCOOCH_3$	69.8 ± 35.9	69.9 ± 36.0	80.0 ± 41.2	73 ± 38	107 ± 55	162 ± 84
Tentative-detection						
t-HCOOH	2.0 ± 1.9	2.4 ± 2.2	2.9 ± 2.7	2.1 ± 1.9	3.6 ± 3.4	5.8 ± 5.5
Upper limit						
CH_3CHO	< 0.62	< 0.75	< 0.95	< 0.65	< 1.2	< 1.9
$(CH_3)_2O$	< 116.1	< 119.6	< 135.0	< 121.6	< 183.8	< 273.6

[a]Local thermodynamic equilibrium (LTE) condition is assumed. N(H_2) is derived from the 1.2 mm dust continuum emission (see Sect. 8.7). Errors are derived from three times the root mean square noise of the integrated intensity

1.8 [33] by assuming that κ_ν depends on the wavelength λ with the equation: $\kappa_\nu = 0.1 \times (0.3\ \mathrm{mm}/\lambda)^\beta\ \mathrm{cm}^2\ \mathrm{g}^{-1}$ [2]. With the dust temperature of 70 K and 130 K, the H_2 column density is obtained to be 9.5×10^{23} and 4.9×10^{23} cm^{-2}, respectively. Then, the fractional abundances relative to H_2 are evaluated to be $(1.3 - 3.9) \times 10^{-10}$ and $(7.3 - 16.2) \times 10^{-9}$ for NH_2CHO and $HCOOCH_3$, respectively, by assuming that the gas temperature is the same as the dust temperature (Table 8.3). These fractional abundances of NH_2CHO and $HCOOCH_3$ are comparable to those reported for the hot corinos: 6×10^{-10} and 9×10^{-9} in IRAS 16293–2422 [13], and $(1 - 9) \times 10^{-10}$ and $(2 - 8) \times 10^{-9}$ in B335 [12]. Therefore, L483 indeed harbors a hot corino activity in the closest vicinity of the protostar.

As mentioned in Sect. 8.4, different molecular emission traces different parts in this source. The extended infalling-rotating envelope is traced by the CS emission, and possibly the CCH emission. On the contrary, the compact component concentrated in the vicinity of the protostar, which is likely to be a disk component inside the centrifugal barrier, is traced by the CS, SO, HNCO, NH_2CHO, and $HCOOCH_3$ emissions. Such a chemical change across the centrifugal barrier is reported for some other sources as described in the previous chapters; L1527, TMC–1A, IRAS 16293–2422 Source A, and Source B (Chaps. 4, 6, 7; [27, 28]). In contrast to the L1527 and TMC–1A cases, the CS and SO emissions in L483 are found to be quite abundant in the compact component concentrated to the protostar. This situation is rather similar to the IRAS 16293–2422 Source A case (Sect. 8.5). Thus, the chemical change would be highly dependent on sources. Hence, it is still essential to investigate chemical structures of various protostellar sources at a sub-arcsecond resolution.

8.8 Summary of This Chapter

The Class 0 protostellar core L483 is observed with ALMA in various molecular lines. The major results are summarized as follows:

(1) A chemical differentiation at a 100 au scale is found in L483. The CCH emission traces the outflow cavity wall, and shows a hole around the protostellar position with a radius of $0''.5$ (100 au) in its distribution. The CS emission traces a part of the outflow cavity wall, the extended envelope component, and the compact component concentrated to the protostar. The SO and HNCO emissions are detected only in the compact component.
(2) In spite of the WCCC character of this source, the COMs, NH_2CHO and $HCOOCH_3$, are firmly detected. Their emission is highly concentrated near the protostar. This is the first spatially-resolved observation of the '*hybrid*' chemical character of WCCC and hot corino chemistry.
(3) The kinematic structure of the disk/envelope system is analyzed by the models of an infalling-rotating envelope and a Keplerian disk. The infalling-rotating envelope model with the protostellar mass from 0.1 to 0.2 $M_\odot$ and the radius of the centrifugal barrier from 30 to 200 au reasonably explains the extended part of the observed PV diagrams of the CS emission. The compact component likely traces the Keplerian disk component inside the centrifugal barrier, although the rotation curve is not resolved well. Hence, the above chemical change seems to be occurring across the centrifugal barrier.
(4) The CS emission traces the disk component in addition to the envelope component in L483, although it is thought to be a good tracer of the infalling-rotating envelope in other sources. As well, the SO emission traces the disk component in L483, which highlights the ring-like structure around the centrifugal barrier in L1527 and TMC−1A. These results would originate from the relatively high luminosity of L483.
(5) The outflow structure of L483 is extended along the southeast-northwest direction. Its kinematic structure traced by the CS emission is well explained by a parabolic outflow model. There is a hint that the outflow has a rotation motion.
(6) The SiO emission shows an extension from the protostellar position. It may be related to a possible local outflow-shock occurring near the centrifugal barrier.

References

1. Agúndez M, Cernicharo JM (2015) Discovery of interstellar ketenyl (HCCO), a surprisingly abundant radical. A&A 577:L5
2. Beckwith SVW, Sargent AI, Chini RS, Guesten R (1990) A survey for circumstellar disks around young stellar objects. AJ, 99:924
3. Ceccarelli C (2004) The hot corinos of solar type protostars. In: Johnstone D, Adams FC, Lin DNC, Neufeeld DA, Ostriker EC (eds) Star formation in the interstellar medium: in honor of David Hollenbach, volume 323 of Astronomical society of the pacific conference series, p 195

4. Chapman NL, Davidson JA, Goldsmith PF, Houde M, Kwon W, Li Z-Y, Looney LW, Matthews B, Matthews TG, Novak G, Peng R, Vaillancourt JE, Volgenau NH (2013) Alignment between flattened protostellar infall envelopes and ambient magnetic fields. ApJ 770(2):151
5. Crimier N, Ceccarelli C, Maret S, Bottinelli S, Caux E, Kahane C, Lis DC, Olofsson J (2010) The solar type protostar IRAS16293-2422: new constraints on the physical structure. A&A 519:A65
6. Favre C, Jørgensen JK, Field D, Brinch C, Bisschop SE, Bourke TL, Hogerheijde MR, Frieswijk WW (2014) Dynamical structure of the inner 100 AU of the deeply embedded protostar IRAS 16293-2422. ApJ, 790(1):55
7. Fuller GA, Lada EA, Masson CR, Myers PC (1995) The infrared nebula and outflow in Lynds 483. ApJ 453:754
8. Green JD, Evans NJ, Jørgensen JK, Herczeg GJ, Kristensen LE, Lee JE, Dionatos O, Yildiz UA, Salyk C, Meeus G, Bouwman J, Visser R, Bergin EA, van Dishoeck EF, Rascati MR, Karska A, van Kempen TA, Dunham MM, Lindberg JE, Fedele D, DIGIT Team (2013) Embedded protostars in the dust, ice, and gas in time (DIGIT) Herschel key program: continuum seds, and an inventory of characteristic far-infrared lines from PACS spectroscopy. ApJ, 770(2):123
9. Hatchell J, Fuller GA, Ladd EF (1999) Temperature predictions for protostellar outflows. A&A 344:687–695
10. Hirota T, Ohishi M, Yamamoto S (2009) A search for carbon-chain-rich cores in dark clouds. ApJ 699(1):585–602
11. Hirota T, Sakai N, Yamamoto S (2010) Depletion of CCS in a candidate warm-carbon-chain-chemistry source L483. ApJ 720(2):1370–1373
12. Imai M, Sakai N, Oya Y, López-Sepulcre A, Watanabe Y, Ceccarelli C, Lefloch B, Caux E, Vastel C, Kahane C, Sakai T, Hirota T, Aikawa Y, Yamamoto S (2016) Discovery of a hot corino in the bok globule B335. ApJ 830(2):L37
13. Jaber AA, Ceccarelli C, Kahane C, Caux E (2014) The census of complex organic molecules in the solar-type protostar IRAS16293-2422. ApJ 791(1):29
14. Jørgensen JK (2004) Imaging chemical differentiation around the low-mass protostar L483-mm. A&A 424:589–601
15. Jørgensen JK, Schöier FL, van Dishoeck EF (2002) Physical structure and CO abundance of low-mass protostellar envelopes. A&A 389:908–930
16. Lee C-F, Mundy LG, Reipurth B, Ostriker EC, Stone JM (2000) CO outflows from young stars: confronting the jet and wind models. ApJ 542(2):925–945
17. Leung GY, Lim J, Takakuwa S (2016) Tracing infall and rotation along the outflow cavity walls of the L483 protostellar envelope. ApJ, 833(1):55
18. Maury AJ, Belloche A, André P, Maret S, Gueth F, Codella C, Cabrit S, Testi L, Bontemps S (2014) First results from the CALYPSO IRAM-PdBI survey. II. Resolving the hot corino in the Class 0 protostar NGC 1333-IRAS2A. A&A, 563:L2
19. McElroy D, Walsh C, Markwick AJ, Cordiner MA, Smith K, Millar TJ (2013) The UMIST database for astrochemistry 2012. A&A 550:A36
20. Müller HS, Schlöder F, Stutzki J, Winnewisser G (2005) The cologne database for molecular spectroscopy, CDMS: a useful tool for astronomers and spectroscopists. J Molecular Struct 742(1-3):215–227
21. Oya Y, López-Sepulcre A, Sakai N, Watanabe Y, Higuchi AE, Hirota T, Aikawa Y, Sakai T, Ceccarelli C, Lefloch B, Caux E, Vastel C, Kahane C, Yamamoto S (2019) Sulfur-bearing species tracing the disk/envelope system in the class I protostellar source Elias 29. ApJ 881(2):112
22. Oya Y, Watanabe Y, López-Sepulcre A, Ceccarelli C, Lefloch B, Favre C, Yamamoto S (2021) Rotating motion of the outflow of IRAS 16293–2422 A1 at its origin point near the protostar. ApJ 921(1):12
23. Park YS, Panis JF, Ohashi N, Choi M, Minh YC (2000) Interferometric observation of the L483 molecular core. ApJ 542(1):344–351
24. Pickett HM, Poynter RL, Cohen EA, Delitsky ML, Pearson JC, Müller HSP (1998) Submillimeter, millimeter and microwave spectral line catalog. J Quant Spec Radiat Transf 60(5):883–890

25. Rice EL, Prato L, McLean IS (2006) An association in the aquila star-forming region: high-resolution infrared spectroscopy of T Tauri stars. ApJ 647(1):432–443
26. Sakai N, Oya Y, Higuchi AE, Aikawa Y, Hanawa T, Ceccarelli C, Lefloch B, López-Sepulcre A, Watanabe Y, Sakai T, Hirota T, Caux E, Vastel C, Kahane C, Yamamoto S (2017) Vertical structure of the transition zone from infalling rotating envelope to disc in the Class 0 protostar, IRAS 04368+2557. MNRAS 467(1):L76–L80
27. Sakai N, Oya Y, López-Sepulcre A, Watanabe Y, Sakai T, Hirota T, Aikawa Y, Ceccarelli C, Lefloch B, Caux E, Vastel C, Kahane C, Yamamoto S (2016) Subarcsecond analysis of the infalling-rotating envelope around the class I protostar IRAS 04365+2535. ApJ 820(2):L34
28. Sakai N, Oya Y, Sakai T, Watanabe Y, Hirota T, Ceccarelli C, Kahane C, Lopez-Sepulcre A, Lefloch B, Vastel C, Bottinelli S, Caux E, Coutens A, Aikawa Y, Takakuwa S, Ohashi N, Yen H-W, Yamamoto S (2014) A chemical view of protostellar-disk formation in L1527. ApJ 791(2):L38
29. Sakai N, Sakai T, Hirota T, Burton M, Yamamoto S (2009) Discovery of the second warm carbon-chain-chemistry source, IRAS15398 - 3359 in lupus. ApJ 697(1):769–786
30. Sakai N, Sakai T, Hirota T, Watanabe Y, Ceccarelli C, Kahane C, Bottinelli S, Caux E, Demyk K, Vastel C, Coutens A, Taquet V, Ohashi N, Takakuwa S, Yen H-W, Aikawa Y, Yamamoto S (2014) Change in the chemical composition of infalling gas forming a disk around a protostar. Nature 507(7490):78–80
31. Sakai N, Yamamoto S (2013) Warm carbon-chain chemistry. Chem Rev 113(12):8981–9015
32. Shirley YL, II Evans NJ, Rawlings JMC, Gregersen EM (2000) Tracing the mass during low-mass star formation. I. Submillimeter continuum observations. ApJS 131(1):249–271
33. Shirley YL, Mason BS, Mangum JG, Bolin DE, Devlin MJ, Dicker SR, Korngut PM (2011) Mustang 3.3 mm continuum observations of Class 0 protostars. AJ, 141(2):39
34. Tafalla M, Myers PC, Mardones D, Bachiller R (2000) L483: a protostar in transition from Class 0 to Class I. A&A 359:967–976
35. Takakuwa S, Kamazaki T, Saito M, Yamaguchi N, Kohno K (2007) ASTE observations of warm gas in low-mass protostellar envelopes: different kinematics between submillimeter and millimeter lines. PASJ 59:1–13
36. Ward-Thompson D, Zylka R, Mezger PG, Sievers AW (2008) Dust emission from star-forming regions. VI. The submillimetre YSO cluster in NGC 2264. A&A, 355:1122–1128
37. Yamamoto S (2017) Introduction to astrochemistry: chemical evolution from interstellar clouds to star and planet formation, Chap. 6
38. Yıldız UA, Kristensen LE, van Dishoeck EF, Hogerheijde MR, Karska A, Belloche A, Endo A, Frieswijk W, Güsten R, van Kempen TA, Leurini S, Nagy Z, Pérez-Beaupuits JP, Risacher C, van der Marel N, van Weeren RJ, Wyrowski F (2015) APEX-CHAMP$^+$ high-J CO observations of low-mass young stellar objects. IV. Mechanical and radiative feedback. A&A 576:A109

Chapter 9
Chemical Differentiation

9.1 Chemical Diversity

9.1.1 Chemical Diversity in a Disk Forming Region

As described in Chap. 1, it is well known that the chemical composition of the protostellar envelopes shows significant diversity at a 1000 au scale [20]. The two distinct cases are the warm carbon-chain chemistry (WCCC) and the hot corino chemistry. In this thesis, the chemical composition and its distribution in the vicinity of a protostar were investigated toward the five representative sources revealing the WCCC and the hot corino chemistry. it was confirmed that the chemical diversity recognized at a few 1000 au scale resolution by single-dish observations can also be seen at a 100 au scale by high angular resolution observations with ALMA (Table 9.1); WCCC sources, L1527 and IRAS 15398−3359 (Chaps. 4 and 5), are rich in unsaturated carbon-chain and related molecules (e.g. CCH and c-C_3H_2) at a few 100 au scale, while they are confirmed to be deficient in saturated complex organic molecules (COMs; e.g. $HCOOCH_3$) even with the high angular resolution and high sensitivity of ALMA. On the other hand, hot corino sources, IRAS 16293−2422 Source A and Source B (Chaps. 6 and 7), are confirmed to be rich in COMs, such as CH_3OH and $HCOOCH_3$, and their distributions are highly concentrated in the vicinity of the protostar at a 100 au scale or smaller. These results suggest that the chemical diversity seen at a few 1000 au scale resolution is indeed seen in a more compact region at a 100 au scale for the above sources as it is.

On the other hand, L483 (Chap. 8), which was recognized as a WCCC candidate source based on single-dish observations [8, 9] shows a different feature. In this source, CCH is indeed abundant at a few 100 au scale with the ALMA observations, which is consistent with its WCCC characteristics. However, some COM lines (HNCO, NH_2CHO, and $HCOOCH_3$), which are specific to the hot corino chemistry, are also detected in the vicinity of the protostar at a 100 au scale or smaller. Since COMs are generally expected to be deficient in WCCC sources as shown in the

Y. Oya, *A Few Tens au Scale Physical and Chemical Structures Around Young Low-Mass Protostars*, Springer Theses, https://doi.org/10.1007/978-981-19-1708-0_9

Table 9.1 Chemical differentiation in low-mass protostellar sources

Source name	Chemistry[a]	Luminosity ($L_\odot$)	Physical structure		
			Envelope	Centrifugal barrier	Disk
L1527[b]	WCCC	1.7	CCH, c-C_3H_2, CS, H_2CO	SO	H_2CO (, CS?)
IRAS 15398−3359[c]	WCCC	1.8	CCH, H_2CO		(H_2CO?)
TMC−1A[d]	WCCC	2.5	CS	SO	
IRAS 16293−2422 Source A[e]	HC	22[f]	OCS, H_2CS	CH_3OH, $HCOOCH_3$	H_2CS (, COMs?)
IRAS 16293−2422 Source B[g]	HC		OCS, H_2CS	CH_3OH, $HCOOCH_3$	(OCS?)
L483[h]	Hybrid	13	CCH, CS	CS, SO	CS, SO, COMs[i]
B335[j]	Hybrid	0.72	CCH, CS	(Not resolved: CS, SO, COMs)[k]	

[a]WCCC stands for the warm carbon-chain chemistry, and HC for the hot corino chemistry
[b]This work (Chap. 4) and [17, 19]. The bolometric luminosity is taken from [5]
[c]This work (Chap. 5). The bolometric luminosity is taken from [12]
[d]Taken from [16]. The bolometric luminosity is taken from [5]
[e]This work (Chap. 6)
[f]Taken from [3]. This value is the sum for Sources A and B
[g]This work (Chap. 7)
[h]This work (Chap. 8). The bolometric luminosity is taken from [21]
[i]Complex organic molecules, such as NH_2CHO and $HCOOCH_3$
[j]Taken from [10]. The bolometric luminosity is taken from [4]
[k]The centrifugal barrier and the disk component are not resolved

above L1527 and IRAS 15398−3359 cases, the detection of the COM emission in L483 was completely unexpected. Hence, L483 shows the WCCC characteristics at the envelope scale (~a few 100 au), while it shows the hot corino activity at the disk formation scale (~100 au). Such a 'hybrid' character source was first identified definitively, although its existence was suggested by single-dish observations of carbon-chain molecules (CCH, C_4H, etc.) and CH_3OH [7, 18].

In this regard, [10] recently reported that the low-mass Class 0 source B335 reveals the hot corino chemistry. B335 is a representative Bok globule [13] and is regarded as one of the best test-bed source for the star formation studies (e.g. [4, 25]). They conducted the ALMA observation toward B335, and detected a number of COM lines (CH_3CHO, NH_2CHO, and $HCOOCH_3$) concentrated within a few 10 au around the protostar. Thus, B335 apparently harbors a hot corino. Nevertheless, they also found that this source shows extended distributions of CCH and c-C_3H_2 over a scale of a few 100 au. Based on these results, B335 can be interpreted as a hybrid character source, as in the case of L483.

In any case, a 100 au scale is just a scale of protoplanetary disks. Thus, all the above results imply that the initial condition in chemical evolution from protoplanetary disks to planets already has diversity at the earliest evolutionary stages.

9.1.2 Which Kind of the Chemical Characteristics is Common?

WCCC and the hot corino chemistry are distinct cases of the chemical diversity of protostellar sources, and have often been thought to be exclusive to each other. The difference in the duration time of the starless core is suggested to be the cause of the difference of the chemical composition [18, 20]. In this picture, the chemical evolution of starless cores from the time when the external UV radiation is shielded to the time of the birth of a protostar is considered. In the initial stage of the chemical evolution, the carbon mostly exists in the ionized form (C^+) or the atomic form (C), and they are gradually converted to CO through gas-phase chemical reactions. The time scale of the conversion (τ_{chem}) is estimated to be a few 10^5 yr. The time scale is determined by the H_3^+ density, which is roughly independent of the H_2 density [22, 24].[1] The chemical time scale is comparable to the dynamical time scale; the free fall time is represented as follows:

$$\tau_{ff} = \sqrt{\frac{3\pi}{32G\rho_0}} \tag{9.1}$$

$$\propto n(H_2)^{-\frac{1}{2}}, \tag{9.2}$$

where G denotes the gravitational constant, and ρ_0 the initial mass density. The time scale of the molecular adsorption onto the dust surface (the depletion time scale) is roughly estimated to be as follows e.g. [24]:

$$\tau_d = \frac{10^9}{n(H_2)} \text{yr} \tag{9.3}$$

$$\propto n(H_2)^{-1}. \tag{9.4}$$

[1] The chemical time scale is determined by formation time scale of molecules. Molecular formation is initiated by the reaction with H_3^+, and hence, the time scale is given by $\frac{1}{k[H_3^+]}$, where k is the Langevin rate coefficient (10^{-9} cm^3 s^{-1} [24]). Since $[H_3^+]$ is 10^4 cm^{-3} independent of the H_2 density, the chemical time scale is about 3×10^5 yr.

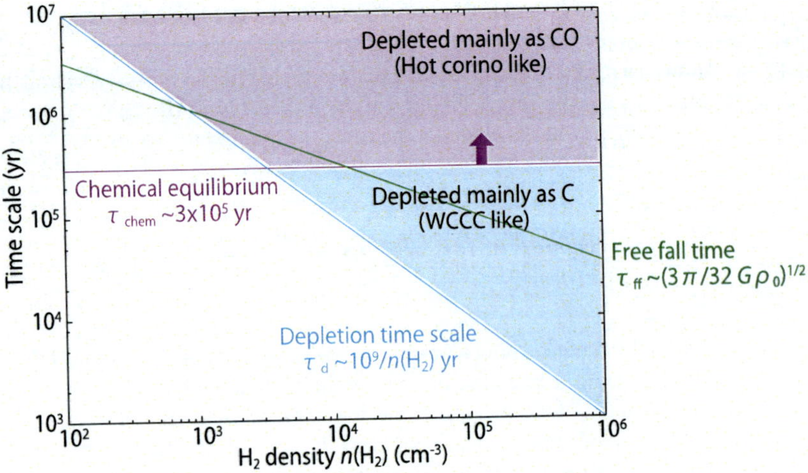

Fig. 9.1 Various time scales as functions of H_2 density. If the depletion time scale is longer than the time scale of the chemical equilibrium, the atomic C is converted to CO, and then CO is adsorbed onto the dust grains. If the depletion time scale is shorter than the time scale of the chemical equilibrium, the atomic C is adsorbed onto the dust grains as it is. This mechanism causes the diversity of the chemical composition of grain mantle, and thereby that of protostellar cores

Figure 9.1 shows the relation between the above time scales and the H_2 density. These time scales ($\sim 10^5$ yr) are comparable to that of the starless-core stage (a few 10^5 yr; Chap. 1).

If the duration time of such a starless core phase is long, the core stays in a moderately dense condition for a long time. Then the gas-phase carbon is mostly converted to CO, and then CO is depleted onto dust grains, as the density increases (Fig. 9.2a). The CO molecules on dust grains are subject to hydrogenation reactions with the H atom to form CH_3OH and probably COMs such as CH_3CHO, $HCOOCH_3$, and $(CH_3)_2O$. After the onset of star formation, dust grains are heated up, and COMs are evaporated. This is the hot corino chemistry. If the duration time is, in contrast, short, the C atom is mainly depleted onto dust grains before it is converted to CO in the gas phase (Fig. 9.2b). The C atoms on dust grains are hydrogenated to form CH_4. After the onset of star formation, CH_4 comes out of dust grains, and produces various carbon-chain molecules in the gas phase as:

$$C^+ + CH_4 \rightarrow C_2H_3^+ + H, \quad (9.5)$$

$$C_2H_3^+ + e^- \rightarrow C_2H_2 + H, C_2H + H + H. \quad (9.6)$$

This is the WCCC. The basic picture of these processes are confirmed by the chemical model calculations [1, 6]. The exclusive nature of WCCC and hot corino chemistry can thus be explained. At the same time, the existence of a source with a hybrid (or

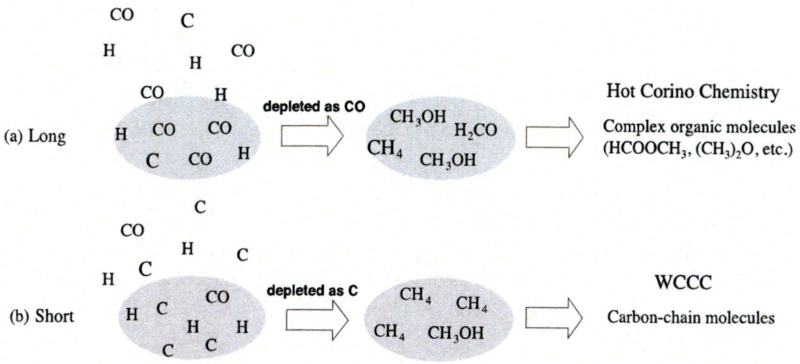

Fig. 9.2 Schematic illustration of the depletion of CO and atomic C onto dust grains and their conversion into COMs and CH_4. (**a**) Long and (**b**) short duration times favor the hot corino chemistry and WCCC, respectively. Taken from [18]. © AAS. Reproduced with permission

intermediate) chemical characteristics is also expected in this scheme, because the depletion of the mixture of C and CO can occur for an appropriate duration time.

Both the WCCC and the hot corino chemistry are triggered by evaporation of molecular species from dust grains. However, the temperature necessary for these two processes are different. The WCCC is triggered by evaporation of CH_4, which occurs in a warm region above 25–30 K. On the other hand, the hot corino chemistry is triggered by evaporation of COMs or disruption of ice mantle of dust grains, which occurs in a hot region above 100 K. Hence, the WCCC region is generally larger than the hot corino. Before the ALMA era, the both regions are sufficiently smaller than the observation beams, and hence, they are just regarded as the chemical feature of protostellar cores. However, these two regions are now resolved at a high angular resolution with ALMA. In fact, the above L483 case seems to be a good example of a source with the hybrid characteristics.

In fact, the hybrid chemical characteristics is consistent with the chemical model calculations. Figure 9.3 shows the result of the chemical model calculation for a contracting protostellar core [1]. In this model, CH_4 is evaporated at the radius of 1000 au from the protostar, and triggers the WCCC. Indeed, the abundances of carbon-chain molecules start to be enhanced inward of the evaporation radius of CH_4. On the other hand, COMs appears in the inner region ($<$100 au), whose temperature is higher than 100 K. Hence, the hybrid chemical characteristics, which is identified in L483 with ALMA, may represent a 'standard' case of the chemical composition of protostellar cores. The recent result for B335 also supports this idea. So far, the representative WCCC sources, L1527 and IRAS 15398−3359, and the representative hot corino source, IRAS 16293−2422, have extensively been studied because of their peculiar chemical nature. However, they should be regarded as the extreme cases, and the hybrid (or intermediate) character sources, such as L483 and B335, would be

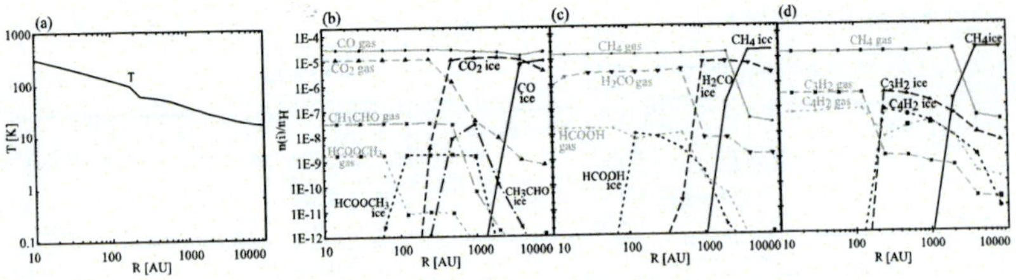

Fig. 9.3 Radial distribution of **a** the gas temperature and **b–d** those of molecules in the protostellar core at the final stage of the chemical model of the contracting cloud. Black lines represent ice-mantle species, while gray lines represent gas-phase species. Taken from [1] with some modifications. © AAS. Reproduced with permission

more common. In order to assess this possibility, statistical studies are required. Such studies will also provide us an important clue to understanding the 'environmental factor' responsible for the appearance of the distinct cases.

9.2 Chemical Change

9.2.1 *Drastic Chemical Change Around the Centrifugal Barrier*

In this thesis, chemical changes across the centrifugal barrier have been found in all the observed protostellar sources. As shown in Fig. 9.4, different molecules trace different physical components around the protostar.

In general, the chemical composition of the gas is changed by gas-phase reactions and/or gas-grain interactions (i.e., evaporation of molecules from dust grains as well as depletion of molecules onto dust grains). For instance, CCH is thought to be broken up by the gas-phase reaction, such as:

$$\mathrm{CCH} + \mathrm{O} \rightarrow \mathrm{CH} + \mathrm{CO}, \tag{9.7}$$

$$\mathrm{He}^+ + \mathrm{CCH} \rightarrow \mathrm{C}^+ + \mathrm{CH} + \mathrm{He}, \tag{9.8}$$

$$\mathrm{H}^+ + \mathrm{CCH} \rightarrow \mathrm{C}_2^+ + \mathrm{H}_2. \tag{9.9}$$

Alternatively, CCH would be depleted onto dust grains in the mid-plane just inside the centrifugal barrier, if the dust temperature is lower than its desorption temperature of about 40 K. The time scale of the depletion of molecules onto dust grains is given as: $t_{\mathrm{dep}} \sim 10^9/n(\mathrm{H}_2)$ yr. Since the mid-plane density just inside the centrifugal barrier is $10^7 - 10^8$ cm^{-3} and the temperature is about 30 K for L1527 [19], the depletion time scale is shorter than the rotation period at the centrifugal barrier ($\sim 10^3$ yr). Thus, CCH is lost in the gas-phase while it goes around the protostar at the centrifugal

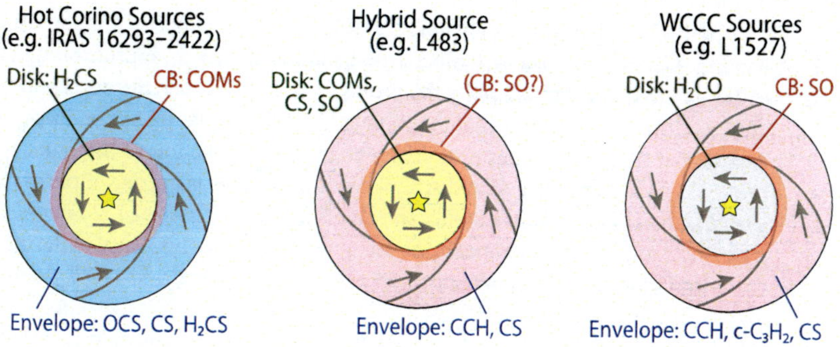

Fig. 9.4 Schematic illustration of the major tracers of the envelope component, the centrifugal barrier ('CB'), and the disk component in hot corino sources (left), WCCC sources (right), and the hybrid case of them (center)

barrier, and it would not be delivered into the gas phase of the disk component inside the centrifugal barrier. In fact, CCH traces the envelope component, but not the disk component in WCCC sources in Table 9.1. Note that CCH may be delivered to the disk component in the solid phase. However, it will readily be polymerized, and never appears in the gas phase.

A similar situation may occur for CS. CS is generally abundant in the gas phase of dense molecular clouds e.g. [23], and hence, they also exist in the infalling-rotating envelope. If the mid-plane temperature is lower than the desorption temperature of CS (35 K; See Appendix A), the gas-phase CS molecules falling from the envelope will be depleted onto dust grains while they are rotating (or stagnated) around the centrifugal barrier. CS is also destroyed by the ionic species like He^+ and H^+. Note that, in contrast to the CCH case, the CS + O reaction has the activation energy of 783 K, so that it is only effective at higher temperature. As a result, CS mainly traces the infalling-rotating envelope in L1527. If the mid-plane temperature is higher than the desorption temperature of CS, CS is not frozen out onto dust grains and can survive in the gas phase. This situation seems to be seen in L483, because the luminosity of L483 (13 $L_\odot$ [3]) is higher than that of L1527 (1.7 $L_\odot$ [5]), as discussed in Chap. 8.

The chemical composition can also be changed by evaporation of molecules from dust grains. There are two mechanisms that liberate surface molecules into the gas phase; those are shock heating and protostellar heating. In the infalling-rotating envelope, the gas is stagnated in front of the centrifugal barrier, and the accreting gas with the infall velocity of a few km s^{-1} causes a weak accretion shock [2, 15]. Then the temperature is raised near the shock front, which evaporates the surface molecules. The SO molecules in L1527 seem to be evaporated from dust grains in this mechanism in front of the centrifugal barrier. Indeed, the temperature is found to be raised up to 60 K or higher around the centrifugal barrier [15, 17]. In IRAS 16293–2422 Source A, a similar enhancement of the temperature around the centrifugal barrier is observed, as described in Chap. 6. This will cause the evaporation of COMs there, which would be the origin of the hot corino chemistry.

Another mechanism responsible for desorption of molecules is the protostellar heating. Generally, the dust temperature becomes higher as approaching to the protostar due to the illumination by the protostar [11, 14]. Since the desorption temperature is different from molecule to molecule, as described in Appendix A, one may expect that this causes chemical differentiation as a function of the radius from the protostar. However, the actual distribution of molecules is not as simple as expected. The different distributions of OCS and H_2CS in IRAS 16293−2422 Source A, which have similar desorption temperature, are not explained in this way. On the other hand, the contribution of this mechanism seems important for CS in L483, as mentioned above. In addition, the protostellar heating in combination with the geometrical effect of the centrifugal barrier may also play an important role of the distribution of COMs in IRAS 16293−2422 Source A, as discussed in Chap. 6. Although the contribution of the protostellar heating to desorption of molecules cannot be ruled out, the physical changes around the centrifugal barrier plays an important role in the chemical differentiation.

9.2.2 Tracers in WCCC and Hot Corino Sources

As demonstrated in Chaps. 4–8, the chemical compositions are useful to disentangle the physical components in the disk-forming regions. For such chemical diagnostics, it is essential to understand what kind of molecular species trace which part of the disk/envelope system.

As shown in Table 9.1 and Fig. 9.4, the chemical changes are seen in different molecules in different sources. The envelope components are traced by unsaturated organic molecules (i.e. carbon-chain and related molecules) in the WCCC sources (L1527, IRAS 15398−3359, and TMC−1A), while it is traced by OCS and H_2CS in the hot corino sources (IRAS 16293−2422 Source A and Source B). The centrifugal barriers are highlighted by SO in the WCCC sources and by COMs in hot corino sources. The disk components inside the centrifugal barrier is traced by various molecules, such as H_2CO, H_2CS, SO, and COMs, depending on sources. It should be noted that the emitting region of one molecule is also different from source to source, as shown in the CS case.

Based on the above results, useful tracers of the disk/envelope system seem to depend on the chemical characteristics of sources, although some trends can be recognized. Thus, confirmation of the applicabilities of the molecules listed in Table 9.1 as the tracers of the physical components in various sources is essential to the establishment of chemical diagnostics for the disk-forming regions. Moreover, characterizing the chemical behavior of more molecular species is also awaited.

References

1. Aikawa Y, Wakelam V, Garrod RT, Herbst E (2008) Molecular evolution and star formation: from prestellar cores to protostellar cores. ApJ 674(2):984–996
2. Aota T, Inoue T, Aikawa Y (2015) Evaporation of grain-surface species by shock waves in a protoplanetary disk. ApJ 799(2):141
3. Crimier N, Ceccarelli C, Maret S, Bottinelli S, Caux E, Kahane C, Lis DC, Olofsson J (2010) The solar type protostar IRAS16293-2422: new constraints on the physical structure. A&A 519:A65
4. II Evans NJ, Di Francesco J, Lee JE, Jørgensen JK, Choi M, Myers PC, Mardones D (2015) Detection of infall in the protostar B335 with ALMA. ApJ 814(1):22
5. Green JD, Evans NJ, Jørgensen JK, Herczeg GJ, Kristensen LE, Lee JE, Dionatos O, Yildiz UA, Salyk C, Meeus G, Bouwman J, Visser R, Bergin EA, van Dishoeck EF, Rascati MR, Karska A, van Kempen TA, Dunham MM, Lindberg JE, Fedele D, DIGIT Team (2013) Embedded protostars in the dust, ice, and gas in time (DIGIT) Herschel key program: continuum SEDs, and an inventory of characteristic far-infrared lines from PACS spectroscopy. ApJ 770(2):123
6. Hassel GE, Herbst E, Garrod RT (2008) Modeling the lukewarm corino phase: is L1527 unique? ApJ 681(2):1385–1395
7. Higuchi AE, Sakai N, Watanabe Y, López-Sepulcre A, Yoshida K, Oya Y, Imai M, Zhang Y, Ceccarelli C, Lefloch B, Codella C, Bachiller R, Hirota T, Sakai T, Yamamoto S (2018) chemical survey toward young stellar objects in the perseus molecular cloud complex. ApJS 236(2):52
8. Hirota T, Ohishi M, Yamamoto S (2009) A search for carbon-chain-rich cores in dark clouds. ApJ 699(1):585–602
9. Hirota T, Sakai N, Yamamoto S (2010) Depletion of CCS in a candidate warm-carbon-chain-chemistry source L483. ApJ 720(2):1370–1373
10. Imai M, Sakai N, Oya Y, López-Sepulcre A, Watanabe Y, Ceccarelli C, Lefloch B, Caux E, Vastel C, Kahane C, Sakai T, Hirota T, Aikawa Y, Yamamoto S (2016) Discovery of a hot corino in the bok globule B335. ApJ 830(2):L37
11. Jørgensen JK, Schöier FL, van Dishoeck EF (2002) Physical structure and CO abundance of low-mass protostellar envelopes. A&A 389:908–930
12. Jørgensen JK, Visser R, Sakai N, Bergin EA, Brinch C, Harsono D, Lindberg JE, van Dishoeck EF, Yamamoto S, Bisschop SE, Persson MV (2013) A recent accretion burst in the low-mass protostar IRAS 15398–3359: ALMA imaging of its related chemistry. ApJ 779(2):L22
13. Keene J, Hildebrand RH, Whitcomb SE, Harper DA (1980) Far-infrared observations of the globule B335. ApJ 240:L43–L46
14. Masunaga H, Miyama SM, Inutsuka S-I (1998) A radiation hydrodynamic model for protostellar collapse. I. The first collapse. ApJ 495(1):346–369
15. Sakai N, Oya Y, Higuchi AE, Aikawa Y, Hanawa T, Ceccarelli C, Lefloch B, López-Sepulcre A, Watanabe Y, Sakai T, Hirota T, Caux E, Vastel C, Kahane C, Yamamoto S (2017) Vertical structure of the transition zone from infalling rotating envelope to disc in the Class 0 protostar, IRAS 04368+2557. MNRAS 467(1):L76–L80
16. Sakai N, Oya Y, López-Sepulcre A, Watanabe Y, Sakai T, Hirota T, Aikawa Y, Ceccarelli C, Lefloch B, Caux E, Vastel C, Kahane C, Yamamoto S (2016) Subarcsecond analysis of the infalling-rotating envelope around the class I protostar IRAS 04365+2535. ApJ 820(2):L34
17. Sakai N, Oya Y, Sakai T, Watanabe Y, Hirota T, Ceccarelli C, Kahane C, Lopez-Sepulcre A, Lefloch B, Vastel C, Bottinelli S, Caux E, Coutens A, Aikawa Y, Takakuwa S, Ohashi N, Yen H-W, Yamamoto S (2014) A chemical view of protostellar-disk formation in L1527. ApJ 791(2):L38
18. Sakai N, Sakai T, Hirota T, Burton M, Yamamoto S (2009) Discovery of the second warm carbon-chain-chemistry source, IRAS15398 - 3359 in lupus. ApJ 697(1):769–786

19. Sakai N, Sakai T, Hirota T, Watanabe Y, Ceccarelli C, Kahane C, Bottinelli S, Caux E, Demyk K, Vastel C, Coutens A, Taquet V, Ohashi N, Takakuwa S, Yen H-W, Aikawa Y, Yamamoto S (2014) Change in the chemical composition of infalling gas forming a disk around a protostar. Nature 507(7490):78–80
20. Sakai N, Yamamoto S (2013) Warm carbon-chain chemistry. Chem Rev 113(12):8981–9015
21. Shirley YL, II Evans NJ, Rawlings JMC, Gregersen EM (2000) Tracing the mass during low-mass star formation. I. Submillimeter continuum observations. ApJS 131(1):249–271
22. Suzuki H, Yamamoto S, Ohishi M, Kaifu N, Ishikawa S-I, Hirahara Y, Takano S (1992) A survey of CCS, HC 3N, HC 5N, and NH 3 toward dark cloud cores and their production chemistry. ApJ 392:551
23. Tatematsu KI, Umemoto T, Kameya O, Hirano N, Hasegawa T, Hayashi M, Iwata T, Kaifu N, Mikami H, Murata Y, Nakano M (1993) Molecular cloud cores in the Orion a cloud. I. Nobeyama CS (1–0) survey. ApJ 404:643
24. Yamamoto S (2017) Introduction to astrochemistry: chemical evolution from interstellar clouds to star and planet formation, Chap. 6
25. Yen H-W, Takakuwa S, Koch PM, Aso Y, Koyamatsu S, Krasnopolsky R, Ohashi N (2015) No Keplerian Disk $>$10 AU around the protostar B335: magnetic braking or young age? ApJ 812(2):129

Chapter 10
Physical Diversity

10.1 Evolution from Envelopes to Disks

The existence of the centrifugal barrier of the infalling-rotating envelope was first reported by [21]. An existence of the centrifugal barrier is naturally derived from simple assumptions of the energy and angular momentum conservation. Thus, it is expected that the centrifugal barrier of the infalling-rotating envelope commonly exists in protostellar sources. In fact, its existence or hint was observationally confirmed in the five young low-mass protostellar sources in this thesis. Moreover, disk structures seem to be formed inside the centrifugal barriers even at early evolutionary stages. The centrifugal barrier seems to be a boundary between the infalling-rotating envelope outside it and the disk component inside it. Hence, the centrifugal barrier will have an important role in the physical mechanism of the disk formation as well as in the chemical evolution described in Chap. 9.

Generally, the size of a rotationally-supported disk, or a Keplerian disk, is thought to correspond to the 'centrifugal radius' [4]. It is the radius where the centrifugal force of the gas and the gravity under the protostellar mass are balanced, and is twice the radius of the centrifugal barrier (Chap. 3). Thus the gas can stably continue rotating around the protostar at the centrifugal radius. As mentioned later, the angular momentum of the gas infalling at a later time tends to be larger, which makes the centrifugal radius larger. This results in the smooth growth of the disk. However, this is a rather static picture. In an infalling-rotating envelope, the rotation speed and the infall speed are the same at the centrifugal radius. Since the rotation speed is the same as the Keplerian speed, the rotation motion can be continuous from the infalling-rotating envelope to the Keplerian disk at the centrifugal radius. On the other hand, there remains the infall motion, and the infalling gas tends to go further inward

A part of this chapter has been published in Oya et al., 2018, ApJ, 863, 72. © AAS. Reproduced with permission.

Y. Oya, *A Few Tens au Scale Physical and Chemical Structures Around Young Low-Mass Protostars*, Springer Theses, https://doi.org/10.1007/978-981-19-1708-0_10

of the centrifugal radius. Indeed, the observational results in Chaps. 4–8 indicate that the gas apparently keeps falling beyond the centrifugal radius toward the centrifugal barrier.

In more realistic cases, the infall motion of the gas will be suppressed due to the gas pressure. But this effect is not effective near the centrifugal radius, as discussed below. The static ($P_{\rm stat}$) and dynamic ($P_{\rm dyn}$) pressures of the gas are represented as follows:

$$P_{\rm stat} = \rho c_s^2, \tag{10.1}$$

$$P_{\rm dyn} = \frac{1}{2}\rho v_{\rm fall}^2, \tag{10.2}$$

where ρ, c_s, and $v_{\rm fall}$ denote the density, the sound speed, and the radial velocity of the gas, respectively. Thus, if ρ can be assumed to be locally constant, the dynamic pressure is larger than the static pressure where $v_{\rm fall}$ is larger than $\sqrt{2}c_s$. Therefore the gas cannot be supported by the static pressure and keeps falling toward the centrifugal barrier.

As for the L1527 case, such a condition is expected until the gas reaches near the centrifugal barrier, as shown in Fig. 10.1. In fact, the static and dynamic pressures are evaluated to be 0.5×10^{-6} and 1.6×10^{-6} dyn cm^{-2} at the centrifugal radius, respectively, where the $v_{\rm fall}$ is 0.9 km s^{-1}. Here, the gas number density is assumed to be 10^8 cm^{-3}, and the averaged particle mass to be$\left(3.83 \times 10^{-24}\right)$ g. c_s is proportional to $\sqrt{T}$, where T is the gas temperature. It is calculated to be 0.35 km s^{-1} for the gas temperature of 30 K in the infalling-rotating envelope of L1527 [21].

The magnetic pressure is represented as:

$$P_M = \frac{B^2}{2\mu_0}, \tag{10.3}$$

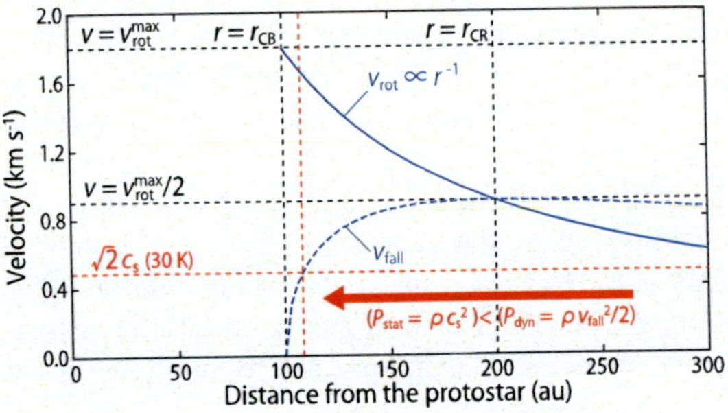

Fig. 10.1 The velocity profile in the infalling-rotating envelope model for the L1527 case. In the infalling-rotating envelope model, the infall velocity is maximum ($v_{\rm fall} = 0.9$ km s^{-1}) at the centrifugal radius ($r_{\rm CR}$ = 200 au). The sound speed c_s is assumed to be 0.35 km s^{-1}, which corresponds to the value at the temperature of 30 K

where B is the magnetic field and μ_0 the magnetic permeability. Then, the following practical expression is obtained:

$$P_M = 3.98 \times 10^{-14} \, B \, (\mu G)^2 \, \text{dyn cm}^{-2}. \tag{10.4}$$

Hence, the magnetic pressure can be comparable to the dynamic pressure, only when the magnetic field is of the order of mG. According to the previous magnetic field measurements using the Zeeman effect, such a strong magnetic field can only be seen in maser spots [3], and is unlikely for the infalling-rotating envelope of the low-mass star forming regions. Furthermore, the ratio of the magnetic pressure to the static gas pressure is assumed to be unity or less in the magnetohydrodynamics (MHD) calculations (e.g. [10]). Otherwise the disk/envelope structure is unstable [8, 24]. Therefore, the magnetic breaking is not expected to be effective around the centrifugal radius in the low-mass star forming regions, neither.

Since ρ and c_s will be increased around the centrifugal barrier due to the stagnation of the gas and the relatively high temperature at the centrifugal barrier (60 K in L1527) to the infalling-rotating envelope (30 K in L1527), the static pressure will be enhanced near the centrifugal barrier. Thus the gas may stop falling before it reaches at the centrifugal barrier. This means that the edge of the rotationally supported disk is smaller than the centrifugal radius.

On the other hand, some cases could be expected, where the infalling-rotating envelope model is not appropriate: e.g., sources with a very small or high protostellar mass. With a small protostellar mass, the infall velocity is small and thus the dynamic pressure of the gas is small. This will result in the static pressure around the centrifugal radius higher than the dynamic pressure. With a high protostellar mass, the gas temperature around the centrifugal radius will be high due to the high luminosity. Thus, the static pressure, which is proportional to T through c_s, could be higher than the dynamic pressure. In these situations, the infall gas may be supported by the static gas and cannot fall toward the centrifugal barrier. In addition, as for the more evolved sources, the infalling envelope gas may be exhausted, and thus the dynamic pressure may not be high enough to push the stagnated gas in front of the centrifugal barrier any more. Therefore, the edge of the disk would be extended to the centrifugal radius.

In contrast, the gas can never go inward of the centrifugal barrier unless it loses the angular momentum and the energy, and hence, it will be stagnated in front of the centrifugal barrier by colliding with the gas infalling afterwards. Such gas stagnation has recently been observed by high-resolution observations of L1527 with ALMA [18] (See Chap. 11). The above mechanism will cause a weak accretion shock around the centrifugal barrier, as mentioned in Chap. 9, which is indeed highlighted by the emission of SO and complex organic molecules (COMs) for the warm carbon-chain chemistry (WCCC) and hot corino sources, respectively. Thus the disk formation, that is the transition from the infalling-rotating envelope to the disk component, is not a straightforward process, but involves discontinuous physical processes.

Although the infalling-rotating envelope model can reasonably explain the basic kinematic structure observed for the five protostellar sources, some caveats need to be mentioned. The infalling-rotating envelope model used in this thesis is a quite simplified one; it does not consider various effects, such as gas pressure, magnetic breaking, self gravity, and time variation of the angular momentum. In reality, these effects will affect *where the infall motion of the gas practically vanishes*. These complex situations considered, it is still unknown *what is occurring between the centrifugal radius and the centrifugal barrier*. Thus, more detailed investigations of the physical structure in the disk forming region are needed to understand the detailed physical mechanisms.

10.2 Angular Momentum of the Envelope Gas

As shown in Chaps. 4–8, the radius of the centrifugal barrier (r_{CB}) varies from source to source; the centrifugal barrier is spatially resolved in some sources (e.g. L1527), while it is not in some other sources in the present study (e.g. IRAS 15398−3359). This suggests that the magnitude ratio of the rotation motion relative to the infall motion depends on sources. Since a disk structure is supported by a rotation motion of the gas around the protostar, characterization of the ratio is essential to understanding the disk formation. To discuss it more quantitatively, the specific angular momentum (j) of the envelope gas is derived. In the infalling-rotating envelope model (Chap. 3), the radius of the centrifugal barrier can be represented in terms of j in the following equation:

$$r_{CB} = \frac{j^2}{2GM}, \tag{10.5}$$

where G denotes the gravitational constant, and M the protostellar mass. Since r_{CB} is proportional to the square of j, a small change in j results in a large change in r_{CB}. Thus, r_{CB} is a good barometer of j. With the Eq. (10.5), j can be derived from M and r_{CB}, which are determined from the observations with the aid of the infalling-rotating envelope model. The j value derived for each source is summarized in Table 10.1. The table includes the results reported for TMC−1A and B335 [6, 19]. In Table 10.1, j seems to vary by an order of magnitude from source to source.

It is theoretically expected that the specific angular momentum increases along the protostellar evolution for a single source. Considering a turbulent parent core, the most fundamental rotation motion is the rigid rotation. In this case, the gas more distant from the center possesses a larger specific angular momentum. Along with the evolution of the source, the gas that was originally more distant from the protostar falls toward the protostar, and it brings a larger specific angular momentum into the vicinity of the protostar. Thus, the specific angular momentum of the infalling gas in the vicinity of the protostar is thought to increase gradually. Since the evolution of the specific angular momentum will affect the physical condition of a future disk structure, its characterization is essential in the study of the disk formation. However, such a trend cannot clearly be seen in Table 10.1. For instance, the j

Table 10.1 Physical Parameters[a]

Source	Evolutional Stage	Chemistry[b]	Inclination Angle[c] (°)	Protostellar mass ($M_\odot$)	Radius of the CB[d] (au)	j^e (10^{-4} km s^{-1} pc)
L1527[f]	Class 0/I	WCCC	85	0.18 ± 0.05	100 ± 20	8.7 ± 1
TMC−1A[g]	Class I	WCCC	70	0.25 ± 0.05	50 ± 10	7.2 ± 1
IRAS 15398−3359[h]	Class 0	WCCC	70	0.02 ± 0.02	<30	<1.6
IRAS 16293−2422 Source A[i]	Class 0	HC	60	0.75 ± 0.25	50 ± 10	12.5 ± 2
IRAS 16293−2422 Source B[j]	Class 0	HC	5	$0.4^{+0.4}_{-0.2}$	40 ± 10	8.2^{+4}_{-2}
L483[k]	Class 0	Hybrid	80	0.15 ± 0.05	100^{+100}_{-70}	7.9^{+4}_{-3}
B335[l]	Class 0	Hybrid	90	>0.13	<10	~2.9

[a]The protostellar mass and the radius of the centrifugal barrier are derived with the aid of the infalling-rotating envelope model, except for the values for B335. They typically contains uncertainties of about a few 10%. See the analysis sections for each source for more details
[b]WCCC stands for the warm carbon-chain chemistry, and HC for the hot corino chemistry
[c]0° for a face-on configuration. The values for L1527, IRAS 15398−3359, and L483 are derived from the outflow analysis (See Table 10.2)
[d]Centrifugal barrier
[e]Specific angular momentum of the envelope gas derived from the protostellar mass and the radius of the centrifugal barrier
[f]This work (Chap. 4) and [20]
[g]Taken from [19]
[h]This table summarizes the values obtained in this work (Chap. 5; [15]), although the protostellar mass has been updated by [13]
[i]This table summarizes the values obtained in this work (Chap. 6; [14]), although they have been updated by [17]
[j]This work (Chap. 7)
[k]This work (Chap. 8)
[l]Taken from [6]. The specific angular momentum is calculated by assuming M of 0.13 $M_\odot$ and r_{CB} of 10 au, just for reference

value for the Class I source TMC−1A is similar to that of the Class 0 sources. The specific angular momentum can also vary from source to source depending on the initial angular momentum of the parent clouds. This contribution may overwhelm the potential evolution of the specific angular momentum in a single source. [27] compiled the measurements of j at various size scales for various sources, and indicated a significant variation of j at the small radius. Their result is consistent with the above result.

10.3 Relation Between the Envelope and the Outflow

The gas at the centrifugal barrier cannot fall into the disk inside it because of the energy and angular momentum conservation. Thus, some mechanisms to extract the angular momentum of the envelope gas are required to form the disk component and also to make the protostar grow further. Outflows are thought to be the most probable candidate for such a mechanism. If this is the case, outflows would be launched from the centrifugal barrier to bring out the angular momentum. In fact, hints of this picture are seen in this thesis; in IRAS 16293−2422 Source B (Chap. 7), the pole-on outflow lobes traced by SiO show a radial offset near the protostar, and their launching point seems to be near the centrifugal barrier traced by OCS and H_2CS (Fig. 7.18). In L483 (Chap. 8), the SiO emission shows an intensity peak at the position apart from the protostar by 100 au, which corresponds to the radius of the centrifugal barrier (Fig. 8.19). Since SiO is known as a shock tracer (e.g. [11]), the results for the both sources imply a possible shock between the outflowing gas and the infalling envelope gas near the centrifugal barrier.

[18] reported that the envelope gas in L1527 is stagnated at the centrifugal barrier, and that it has a substantial extension perpendicular to the mid-plane. It is likely that a part of the gas is moving away from the mid-plane. [18] suggested a possibility that this outflowing motion forms the so-called ‘disk winds’ or ‘low-velocity molecular outflow’ launched at the centrifugal barrier. If so, one would expect a rotation motion of the outflow particularly in the vicinity of the centrifugal barrier. However, a definitive evidence of the rotation motion of the outflow has not been seen in L1527 (Chap. 4). Although a hint of the outflow rotation is seen in L483 (Chap. 8), the relation between the outflow launching and the centrifugal barrier is still controversial at this stage.

If the centrifugal barrier is responsible for the outflow launch, there would be some relations between the radius of the centrifugal barrier and the outflow shape. For instance, the outflow launched from the smaller radius of the centrifugal barrier may be more collimated due to less centrifugal motion. To inspect such a possibility, the outflow parameter C is plotted against the specific angular momentum of the infalling-rotating envelope (Fig. 10.2). However, a clear relation is not seen between them at the present stage; there is a big difference in C between L1527 and L483, having the similar specific angular momentum. More samples are apparently needed.

10.4 Evolution of Outflows

The C and v_0 values for the several sources are listed in Table 10.2, where the error ranges for L1527, IRAS 15398−3359, and L483 are estimated by eye from the fits with various parameters. Since the observations with ALMA in this thesis are focused on the central region around the protostar, the dynamical time scales cannot be estimated adequately from these observational data. Therefore, the dynamical time scales for L1527 and IRAS 15398−3359 in Table 10.2 are taken from [28].

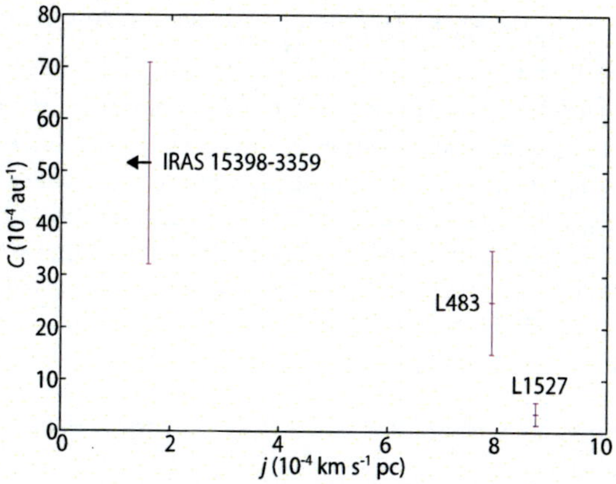

Fig. 10.2 Plots of the curvature parameter (C) of the outflow against the specific angular momentum of the infalling-rotating envelope. The value of the specific angular momentum for IRAS 15398−3359 is the upper limit. Taken from [16].

Table 10.2 Best fit parameters for the models of the outflow cavity walls and the dynamical time scales

Source	Distance (pc)	t_{dyn}[a] (10^3 yr)	Inclination angle[b] (°)	C (10^{-4} au^{-1})	v_0 (10^{-4} km s^{-1})
L1527[c]	137	20.6 (east)[d], 6.5 (west)[d]	85 ± 10	3.6 ± 2.2	7.3 ± 1.5
IRAS 15398−3359[c]	155	0.9 (northeast)[d], 1.8 (southwest)[d]	70 ± 10	52 ± 19	25 ± 2
L483[c]	200	3	80 ± 5	25 ± 10	15 ± 5
VLA 05487[e]	460	~7	71 ± 3	4.8 ± 1.1	7.0 ± 1.1
RNO 91[e]	160	~20	70 ± 4	1.3 ± 0.25	2.4 ± 0.6
L1448C[f]	250	~0.24	69	24, 32	200
HH 46/47[g]	450	9	61 ± 1	6.7 ± 1.1	51 ± 4

[a]Dynamical timescale of an outflow lobe
[b]0° for a pole-on configuration
[c]This work (Chaps. 4, 5, 8)
[d]Determined from the CO ($J = 3 - 2$) emission [28]
[e]Determined from the CO ($J = 1 - 0$) emission [7]
[f]Determined from the CO ($J = 3 - 2$) emission [5]
[g]Determined from the CO ($J = 1 - 0$) emission [1]

10.4.1 Comparison Between L1527 and IRAS 15398−3359

Both of L1527 (Chap. 4) and IRAS 15398−3359 (Chap. 5) have a nearly edge-on configuration of the outflow/envelope structure. However, the appearance of their outflows shown in the integrated intensity maps are quite different from each other. The outflow of L1527 shows a butterfly-feature, while that of IRAS 15398−3359 is well collimated. While an offset is assumed between the two lobes in L1527 to explain the outflow structure as suggested by [25] (Sect. 4.2.3), no such offset is seen for the two lobes in IRAS 15398−3359. It should be noted that the outflow component is contaminated with the envelope component within 1″ from the protostar position in IRAS 15398−3359. Apart from the offset, the difference of the opening angles of the outflow cavity is reflected on the difference of the values of C by a factor of 14 (Table 10.2).

10.4.2 Relation to Dynamical Ages

The outflow parameters are reported for the Class 0/I source VLA 05487 in the Orion dark cloud L1617 and the Class II/III source RNO 91 in the L43 molecular cloud by [7]. Moreover, those for the Class 0 source L1448C (L1448 mm) in Perseus are reported by [5], and those for the HH 46/47 molecular outflow on the outskirts of the Gum Nebula by [1]. The outflow parameters for these sources are summarized in Table 10.2. Here, it should be noted that the coefficients of proportionality reported for these sources are in the unit of arcsecond:

$$C_{\rm as} = CD, \quad v_{\rm as} = v_0 D, \tag{10.6}$$

where D denotes the distance of the source.

Figure 10.3 shows a semi-log plot of C versus dynamical time scale $t_{\rm dyn}$ for these seven sources. The green dashed line in Fig. 10.3 represents the best-fit result, where an equal weight is assumed for all the sources. Although the number of the sources is too small for statistical arguments, the parameter C seems to decrease exponentially with the dynamical time scale of outflows. The correlation coefficient for this plot is −0.95. This feature is consistent with the previous works reporting a relationship between the opening angles of outflows and the source ages (e.g. [2, 22, 26]). The trend of increasing opening angle as increasing age is also revealed by the theoretical simulations (e.g. [9, 12, 23]). However, previous observational studies are based on morphology of outflows. Although the appearance of integrated intensity maps of outflows is affected by the inclination angle, its effect is not considered for some cases. In contrast, the outflow parameters reported in this thesis are derived from both the geometrical and kinematic structures of the outflow by using the parabolic model considering the inclination angle. The results quantitatively supports the trend suggested previously. As described above, the angular momentum of the envelope

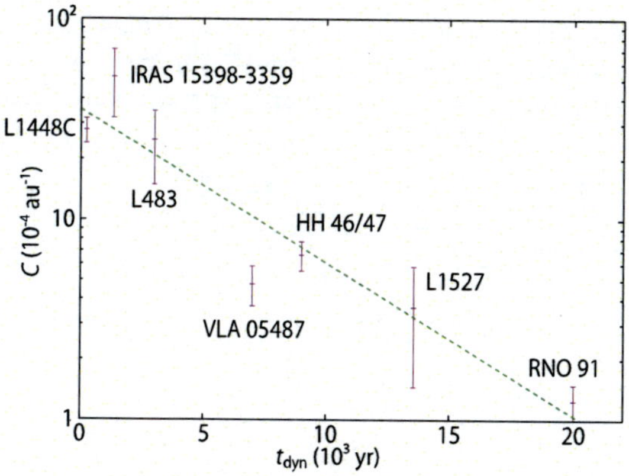

Fig. 10.3 Relation between the curvature parameter (C) of an outflow and its dynamical time scale. The dynamical time scales for L1527 and IRAS 15398−3359 are the averaged value of the two lobes. The green dashed line represents the best-fit function: $\log C = (-0.18 \pm 0.03) \times (t_{\rm dyn} \times 10^{-3}) + (-5.6 \pm 0.3)$, where the uniform weights are applied to all the sources. The correlation coefficient for this plot is −0.95. Taken from [16]. © AAS. Reproduced with permission

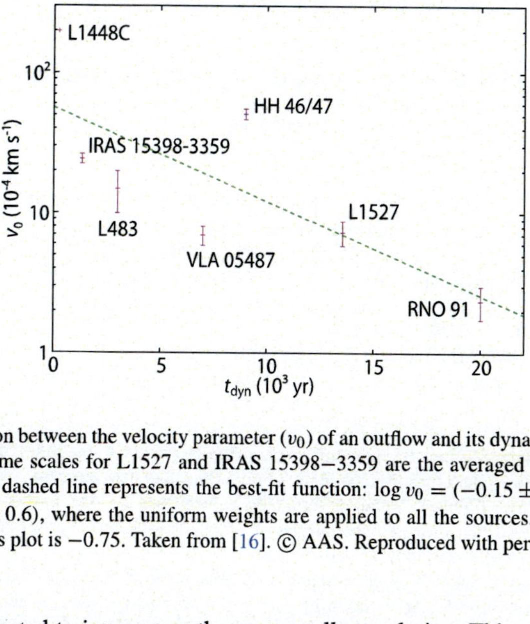

Fig. 10.4 Relation between the velocity parameter (v_0) of an outflow and its dynamical time scale. The dynamical time scales for L1527 and IRAS 15398−3359 are the averaged value of the two lobes. The green dashed line represents the best-fit function: $\log v_0 = (-0.15 \pm 0.06) \times (t_{\rm dyn} \times 10^{-3}) + (-5.2 \pm 0.6)$, where the uniform weights are applied to all the sources. The correlation coefficient for this plot is −0.75. Taken from [16]. © AAS. Reproduced with permission

gas is also expected to increase as the protosetellar evolution. This may contribute to decreasing the C value (Fig. 10.2).

Figure 10.4 shows a semi-log plot of v_0 versus dynamical time scale $t_{\rm dyn}$. Although the correlation coefficient of −0.75 is lower than that for the plot in Fig. 10.3, the velocity of the gas also seems to decrease exponentially with the dynamical time scale of outflows. There seems to reveal a trend that outflows with shorter dynamical time scales have higher v_0. This trend is derived from the velocity structure of the outflow, which was not considered in the morphological studies. Although the parameters derived in this thesis (L1527, IRAS 15398−3359, and L483) are obtained with observations focused on narrow regions without the whole structure of the outflows, they are consistent with the values previously reported based on the analysis of the large-scale outflow structures. Therefore, outflows can be characterized with the aid of the parabolic outflow model focused on a narrow region (a few 100 au scale) around the protostar, even if the whole structure is not observed. This strengthens the idea that the acceleration and geometry of the outflow are mostly defined in a very small region near the protostar.

References

1. Arce HG, Mardones D, Corder SA, Garay G, Noriega-Crespo A, Raga A C (2013) ALMA observations of the HH 46/47 molecular outflow. ApJ 774(1):39
2. Arce HG, Sargent AI (2006) the evolution of outflow-envelope interactions in low-mass protostars. ApJ 646(2):1070–1085
3. Crutcher RM (2012) magnetic fields in molecular clouds. ARA&A 50:29–63
4. Hartmann L (2009) Accretion processes in star formation, 2nd edn
5. Hirano N, Ho PT, Liu S-Y, Shang H, Lee C-F, Bourke TL (2010) Extreme active molecular jets in L1448C. ApJ 717(1):58–73
6. Imai M, Sakai N, Oya Y, López-Sepulcre A, Watanabe Y, Ceccarelli C, Lefloch B, Caux E, Vastel C, Kahane C, Sakai T, Hirota T, Aikawa Y, Yamamoto S (2016) Discovery of a Hot Corino in the Bok Globule B335. ApJ 830(2):L37
7. Lee C-F, Mundy LG, Reipurth B, Ostriker EC, Stone JM (2000) CO outflows from young stars: confronting the jet and wind models. ApJ 542(2):925–945
8. Machida M, Hayashi MR, Matsumoto R (2000) Global simulations of differentially rotating magnetized disks: formation of low-β filaments and structured coronae. ApJ 532(1):L67–L70
9. Machida MN, Hosokawa T (2013) Evolution of protostellar outflow around low-mass protostar. MNRAS 431(2):1719–1744
10. Machida MN, Inutsuka SI, Matsumoto T (2011) Effect of magnetic braking on circumstellar disk formation in a strongly magnetized cloud. PASJ 63:555
11. Mikami H, Umemoto T, Yamamoto S, Saito S (1992) Detection of SiO emission in the L1157 dark cloud. ApJ 392:L87
12. Offner SS, Lee EJ, Goodman AA, Arce H (2011) Radiation-hydrodynamic Simulations of Protostellar Outflows: synthetic observations and data comparisons. ApJ 743(1):91
13. Okoda Y, Oya Y, Sakai N, Watanabe Y, Jørgensen JK, Van Dishoeck EF, Yamamoto S (2018) The co-evolution of disks and stars in embedded stages: the case of the very-low-mass protostar IRAS 15398–3359. ApJ 864(2):L25
14. Oya Y, Sakai N, López-Sepulcre A, Watanabe Y, Ceccarelli C, Lefloch B, Favre Cécile, Yamamoto S (2016) Infalling-rotating motion and associated chemical change in the envelope of IRAS 16293–2422 source A studied with ALMA. ApJ 824(2):88
15. Oya Y, Sakai N, Sakai T, Watanabe Y, Hirota T, Lindberg JE, Bisschop SE, Jørgensen JK, van Dishoeck EF, Yamamoto S (2014) A substellar-mass protostar and its outflow of IRAS 15398–3359 revealed by subarcsecond-resolution observations of H_2CO and CCH. ApJ 795(2):152
16. Oya Y, Sakai N, Watanabe Y, López-Sepulcre A, Ceccarelli C, Lefloch B, Yamamoto S (2018) Sub-arcsecond kinematic structure of the outflow in the vicinity of the protostar in L483. ApJ 863(1):72

17. Oya Y, Yamamoto S (2020) Substructures in the disk-forming region of the class 0 low-mass protostellar source IRAS 16293–2422 source A on a 10 au scale. ApJ 904(2):185
18. Sakai N, Oya Y, Higuchi AE, Aikawa Y, Hanawa T, Ceccarelli C, Lefloch B, López-Sepulcre A, Watanabe Y, Sakai T, Hirota T, Caux E, Vastel C, Kahane C, Yamamoto S (2017) Vertical structure of the transition zone from infalling rotating envelope to disc in the Class 0 protostar, IRAS 04368+2557. MNRAS 467(1):L76–L80
19. Sakai N, Oya Y, López-Sepulcre A, Watanabe Y, Sakai T, Hirota T, Aikawa Y, Ceccarelli C, Lefloch B, Caux E, Vastel C, Kahane C, Yamamoto S (2016) Subarcsecond analysis of the infalling-rotating envelope around the class I protostar IRAS 04365+2535. ApJ 820(2):L34
20. Sakai N, Oya Y, Sakai T, Watanabe Y, Hirota T, Ceccarelli C, Kahane C, Lopez-Sepulcre A, Lefloch B, Vastel C, Bottinelli S, Caux E, Coutens A, Aikawa Y, Takakuwa S, Ohashi N, Yen H-W, Yamamoto S (2014) A chemical view of protostellar-disk formation in L1527. ApJ 791(2):L38
21. Sakai N, Sakai T, Hirota T, Watanabe Y, Ceccarelli C, Kahane C, Bottinelli S, Caux E, Demyk K, Vastel C, Coutens A, Taquet V, Ohashi N, Takakuwa S, Yen H-W, Aikawa Y, Yamamoto S (2014) Change in the chemical composition of infalling gas forming a disk around a protostar. Nature 507(7490):78–80
22. Seale JP, Looney LW (2008) Morphological evolution of bipolar outflows from young stellar objects. ApJ 675(1):427–442
23. Shang H, Allen A, Li Z-Y, Liu C-F, Chou M-Y, Anderson J (2006) A unified model for bipolar outflows from young stars. ApJ 649(2):845–855
24. Shibata K, Tajima T, Matsumoto R (1990) Magnetic accretion disks fall into two types. ApJ 350:295
25. Tobin JJ, Hartmann L, Calvet N, D'Alessio P (2008) Constraining the envelope structure of L1527 IRS: infrared scattered light modeling. ApJ 679(2):1364–1384
26. Velusamy T, Langer WD, Thompson T (2014) HiRes deconvolved spitzer images of 89 protostellar jets and outflows: new data on the evolution of the outflow morphology. ApJ 783(1):6
27. Yen H-W, Koch PM, Takakuwa S, Krasnopolsky R, Ohashi N, Aso Y (2017) Signs of early-stage disk growth revealed with ALMA. ApJ 834(2):178
28. Yıldız UA, Kristensen LE, van Dishoeck EF, Hogerheijde MR, Karska A, Belloche A, Endo A, Frieswijk W, Güsten R, van Kempen TA, Leurini S, Nagy Z, Pérez-Beaupuits JP, Risacher C, van der Marel N, van Weeren RJ, Wyrowski F (2015) APEX-CHAMP$^+$ high-J CO observations of low-mass young stellar objects. IV. Mechanical and radiative feedback. A&A 576:A109

Chapter 11
Conclusion

11.1 Summary of This Thesis

Major outcomes of this thesis are summarized below. The questions raised in Introduction (Chap. 1) have been addressed, at least partly.

(1) The infalling-rotating envelope and its centrifugal barrier are common occurrences in the observed sources. These physical structures are seen regardless of the chemical characteristics (WCCC/hot corino chemistry/hybrid) of the sources. The infalling-rotating envelope model based on the simple physical assumptions (the energy and angular momentum conservation) works well to characterize the observed kinematic structure. Thus, the model seems to capture the basic physics in disk forming regions.

(2) The centrifugal barrier is most likely the transition zone from the infalling-rotating envelope to the disk inside it. The kinematic structure is discontinuously changed across the centrifugal barrier. It is expected that this physical change is also related to the launching mechanism of the outflow as an extraction mechanism of the angular momentum of the gas. Although some hints are found in this thesis, their relation is not obvious at the present stage.

(3) The fact that the observed kinematic structure is well represented by the infalling-rotating envelope model implies that the magnetic field does not effectively affect the gas motion at least in the infalling-rotating envelope at a few 100 au scale.

(4) The chemical change across the centrifugal barrier is seen in all the observed sources, which is probably caused by the accretion shock in front of the centrifugal barrier as well as the protostellar heating. The molecular species tracing each physical component of the disk forming region (the infalling-rotating envelope, the centrifugal barrier, and the disk component) is different depending on the chemical characteristics of the sources. This would result in a chemical diversity in protoplanetary disks and eventually in planets.

Y. Oya, *A Few Tens au Scale Physical and Chemical Structures Around Young Low-Mass Protostars*, Springer Theses, https://doi.org/10.1007/978-981-19-1708-0_11

(5) The chemical processes in the stagnated gas around the centrifugal barrier are expected to determine the chemical heritage passed from the infalling-rotating envelope to the disk component. Thus, the initial condition of the chemical evolution in protoplanetary disks is defined in the transition zone.
(6) Several molecular lines are recognized as good tracers for the physical components. The chemical diagnostics will be a powerful tool for star-formation studies in the ALMA era, although it is still in its infancy.

11.2 Future Prospects

11.2.1 Transition Zone from the Envelope to the Disk

In this thesis, it has been demonstrated that images at a sub-arcsecond resolution unveil unexpected phenomena. However, there are still complex physical components remained to be resolved, for instance, the detailed structure of the transition zone from the infalling-rotating envelope to the disk, the stagnation of the gas at the centrifugal barrier, and the launching points of the outflow. The physical parameters for some sources observed in this thesis are just upper limits. Higher angular resolution observations will allow us to characterize them more accurately.

Based on these motivations, such higher angular-resolution observations are going to be conducted with ALMA. These projects plan to observe the sources with which this thesis has dealt at an angular resolution of up to 0.″07 (10 au). Some parts of these programs have already been executed in these years. With a high angular-resolution observation (0.″2) toward L1527, [5] has indeed confirmed the extension of the scale height of the envelope gas at the centrifugal barrier due to the gas stagnation, as described in Chap. 10. The higher angular-resolution observations will provide us with new insights into the disk formation study.

11.2.2 How about in More Evolved Sources?

The concept of the infalling-rotating envelope and its centrifugal barrier has been confirmed to be common at the evolutionary stages of Class 0 and I. Moreover, the chemical process occurring around the centrifugal barrier would determine the initial condition for the chemical evolution in the disk formation. This thesis shows that chemical diversity is already seen at its earliest stage. Then, the next question naturally rises: how does it evolve in the future? Thanks to the high sensitivity of ALMA, the weak molecular line emission of various molecular species is being detected in protoplanetary disks (e.g. [2]). For instance, the H_2CS line is detected toward MWC 480 (IAU Symposium 332). This detection is interesting, because H_2CS is detected in the disk component of IRAS 16293−2422 Source A (Chap. 6). The chemical characterization of more sources in the Class II stage is urgent to delineate the whole scenario of the chemical evolution from infalling-rotating envelopes to protoplanetary disks and eventually to planets.

11.2.3 Chemical Heritage: Importance of Sulfur Chemistry

Understandings of the behavior of various molecular species in the disk-forming region are essential to building the sound bases of the chemical diagnostics. The observational results in this thesis suggest that S-bearing molecules, such as CS, SO, OCS, and H_2CS, would be key species in the chemical differentiation in the disk forming region. However, their chemical behaviors are still unknown both observationally and theoretically. Since they will be sensitive to physical changes in disk formation, physical characterization of each source is required in order to understand the chemical behaviors of S-bearing molecules. In IRAS 16293−2422 Source A (Chap. 6), H_2CS traces the disk component while OCS does not in spite of their similar desorption temperature (Chap. 9). It suggests that there are some chemical mechanisms which reduces the OCS abundance and enhance the H_2CS abundance inside the centrifugal barrier. Identification of the specific chemical processes responsible for this trend is highly awaited. Such chemical processes will also affect the chemical heritage delivered into the disk. The detection of H_2CS toward the protoplanetary disk mentioned above may further indicate that it is delivered into the planets. Moreover, a recent study with Rosetta also found various S-bearing species in the coma of comet 67P/Churyumov-Gerasimenko [1]. Hence, the sulfur chemistry is the central issue for astrochemical studies of the disk formation in the next decade.

So far, sulfur chemistry has been studied for molecular clouds and star-forming regions (e.g. [3, 4, 6–8]). The most difficult point in the sulfur chemistry is that we do not exactly know the sulfur abundance available for the gas-phase reactions. In the previous models, 99% of sulfur is simply assumed to be depleted onto dust grains. However, this fraction is quite arbitrary. Moreover, we do not exactly know the major form of sulfur on dust grains, although H_2S is thought to be a potential candidate.

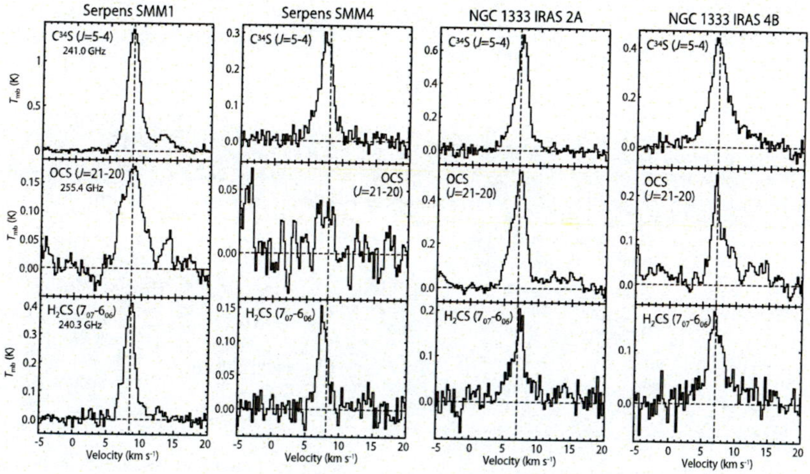

Fig. 11.1 Spectra of S-bearing molecules observed toward the four hot corino sources with the IRAM 30 m telescope. The vertical dashed line represents the systemic velocity of each source. Broad OCS lines, which would come from the vicinity of the protostar, are detected

Hence, the sulfur chemistry seems unpredictable at the present stage in hot regions like the disk forming regions, where extensive gas-grain interaction is occurring. Under such a circumstance, we need to start with understandings of abundances and distributions of various S-bearing molecules in protostellar cores.

Based on this prospect, a spectral line survey observation of S-bearing species has been started with single-dish telescopes (IRAM 30 m, NRO 45 m). The OCS and H_2CS lines were successfully detected at the wavelength of 1.2 mm toward several hot corino sources (Fig. 11.1), such as Serpens SMM1, SMM4, NGC 1333 IRAS 2A, and NGC 1333 IRAS 4B. Although these species has not been popular in the observational studies of low-mass protostellar sources, understandings of their detailed distribution in the vicinity of the protostar will be essential to the establishment of the chemical diagnostics and further exploration of the chemical evolution from molecular clouds to star-forming regions.

The above future projects are schematically illustrated in Fig. 11.2. By tackling these projects with a full use of ALMA, I would like to contribute to unveiling the whole picture of the physical and chemical evolution in formation of Solar-type stars.

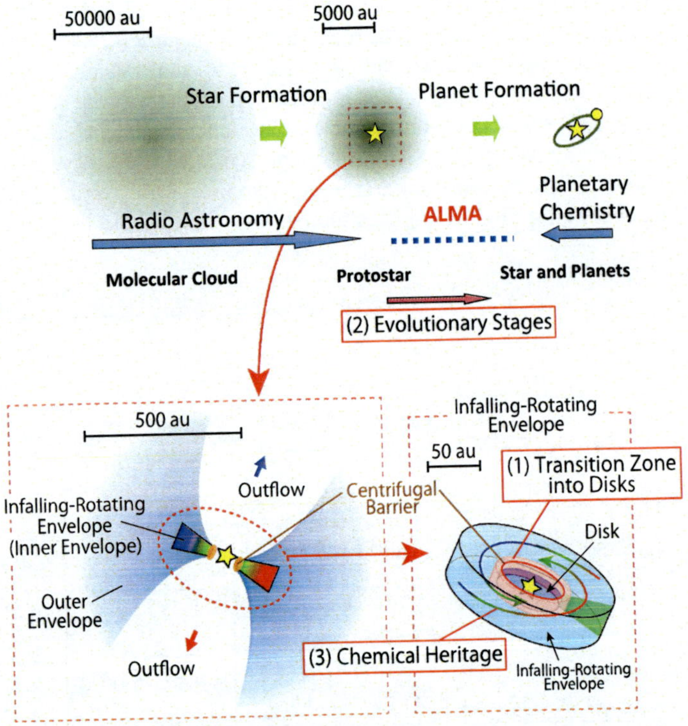

Fig. 11.2 Schematic illustration of the protostellar evolution process (top). The bottom illustrations depict the disk/envelope system of young low-mass protostellar sources, whose physical and chemical structures have been characterized in this thesis. The focusing points of the future projects mentioned in Sect. 11.2 are also shown

References

1. Calmonte U, Altwegg K, Balsiger H, Berthelier JJ, Bieler A, Cessateur G, Dhooghe F, van Dishoeck EF, Fiethe B, Fuselier SA, Gasc S, Gombosi TI, Hässig M, Le Roy L, Rubin M, Sémon T, Tzou CY, Wampfler SF (2016) Sulphur-bearing species in the coma of comet 67P/Churyumov-Gerasimenko. MNRAS 462:S253–S273
2. Huang J, Öberg KI, Qi C, Aikawa Y, Andrews SM, Furuya K, Guzmán VV, Loomis RA, van Dishoeck EF, Wilner DJ (2017) An ALMA survey of DCN/$H^{13}CN$ and DCO^+/$H^{13}CO^+$ in protoplanetary disks. ApJ 835(2):231
3. Millar TJ, Herbst E (1990) Organo-sulphur chemistry in dense interstellar clouds. A&A 231:466–472
4. Prasad SS, Jr. Huntress WT (1982) Sulfur chemistry in dense interstellar clouds. ApJ 260:590–598
5. Sakai N, Oya Y, Higuchi AE, Aikawa Y, Hanawa T, Ceccarelli C, Lefloch B, López-Sepulcre A, Watanabe Y, Sakai T, Hirota T, Caux E, Vastel C, Kahane C, Yamamoto S (2017) Vertical structure of the transition zone from infalling rotating envelope to disc in the Class 0 protostar, IRAS 04368+2557. MNRAS 467(1):L76–L80
6. Wakelam V, Castets A, Ceccarelli C, Lefloch B, Caux E, Pagani L (2004) Sulphur-bearing species in the star forming region L1689N. A&A 413:609–622
7. Wakelam V, Ceccarelli C, Castets A, Lefloch B, Loinard L, Faure A, Schneider N, Benayoun JJ (2005) Sulphur chemistry and molecular shocks: the case of NGC 1333-IRAS 2. A&A 437(1):149–158
8. Wakelam V, Hersant F, Herpin F (2011) Sulfur chemistry: 1D modeling in massive dense cores. A&A 529:A112

Appendix A
Desorption Temperature

In radio observational studies, rotational spectral lines are observed in interstellar clouds. The line emission (or absorption) arises from freely rotating molecules only in the gas phase. However, interstellar molecules also exist on the surface of dust grains and in the ice mantles. In disk forming regions, the dust temperature is high enough for various molecular species to evaporate/sublimate from dust grains, and it cause a substantial change in the chemical composition of the gas. Therefore, the evaporation/sublimation of molecules from dust grains should be considered in the study of the chemical evolution as well as the chemical processes. Here, the '*desoprtion temperatures*' are evaluated for some molecular species observed in this thesis. Desorption temperature of a molecular species (or also called as *'adsorption temperature'*) is a typical temperature at which the molecular species desorbs from dust grains.

A.1 Balance of the Adsorption and the Desorption

When a molecule collides with a dust grain, it can be adsorbed on dust surfaces due to attracting forces, such as van der Waals force and/or electrostatic forces. The rate of the adsorption of molecules is represented as follows [2]:

$$R_{\mathrm{ads}} = n\,(\mathrm{X})\,S\sigma < v >, \tag{A.1}$$

where $n(\mathrm{X})$ denotes the number density of molecule X in the gas phase, σ the geometrical cross section of dust grains, and $< v >$ the average speed of X. S is the sticking probability, with which a molecule is adsorbed in its collision with a dust grain. S of unity is assumed here, which means that a molecules is adsorbed in every collision.

Adsorbed molecules are thermally desorbed from dust surfaces into the gas phase. The rate of the thermal desorption depends on *'desorption energy'* (or *'binding energy'*) E_{des} of each molecular species, and it can be represented as follows [2]:

Y. Oya, *A Few Tens au Scale Physical and Chemical Structures Around Young Low-Mass Protostars*, Springer Theses, https://doi.org/10.1007/978-981-19-1708-0

$$R_{\mathrm{des}} = N_s\left(\mathrm{X}\right)\nu_0 \exp\left(-\frac{E_{\mathrm{des}}}{k_{\mathrm{B}}T}\right), \tag{A.2}$$

where $N_s(\mathrm{X})$ denotes the number of molecule X on a particular dust grain, k_{B} the Boltzmann constant, T the dust temperature. In a chemical picture, the desorption of a molecule is caused by the vibration between the molecule and the dust surface. The characteristic frequency of the vibration mode is denoted as ν_0, which is typically 10^{12} Hz [2]. $N_s(\mathrm{X})$ can be represented as follows:

$$n_s\left(\mathrm{X}\right) = N_s\left(\mathrm{X}\right) n_g, \tag{A.3}$$

where $n_s(\mathrm{X})$ denotes the number density of molecules X on dust grains in a unit volume, and n_g the number density of the dust in a unit volume. Let Σ the effective collision area of dust per H molecule, it is related to n_g as:

$$\Sigma = \frac{n_g \sigma}{n_0}, \tag{A.4}$$

where n_0 denotes the number density of H nuclei: $n_0 = n\left(\mathrm{H}\right) + 2n\left(\mathrm{H_2}\right)$. Thus, the rate of the thermal desorption can be represented as:

$$R_{\mathrm{des}} = \frac{n_s\left(\mathrm{X}\right)\sigma}{n_0 \Sigma}\nu_0 \exp\left(-\frac{E_{\mathrm{des}}}{k_{\mathrm{B}}T}\right). \tag{A.5}$$

In the steady state, the rates of the adsorption and thermal desorption of molecules are balanced. Thus, the following relation is obtained:

$$n\left(\mathrm{X}\right) < v > = \frac{n_s\left(\mathrm{X}\right)}{n_0 \Sigma}\nu_0 \exp\left(-\frac{E_{\mathrm{des}}}{k_{\mathrm{B}}T}\right). \tag{A.6}$$

Here, the desorption temperature is the temperature at which the number density of molecules X in the gas phase equals to that on dust grains in unit volume: $n\left(\mathrm{X}\right) = n_s\left(\mathrm{X}\right)$. Under this condition, the desorption temperature (T_{des}) can be represented as:

$$k_{\mathrm{B}} T_{\mathrm{des}} = E_{\mathrm{des}}\left(\log \frac{\nu_0}{n_0 \Sigma < v >}\right)^{-1}. \tag{A.7}$$

Here, the desorption temperature of molecule X is proportional to its desorption energy. A typical value of Σ is 10^{-22} cm^2 [2].

Figure A.1 shows the relation between the desorption temperature and the desorption energy with typical values for n_0 and $< v >$. From the plots, the factor of proportionality of T_{des} to E_{des} is approximated to be 50–60. Assuming the factor to be 55, the desorption temperatures of some molecular species can be roughly estimated as summarized in Table A.1.

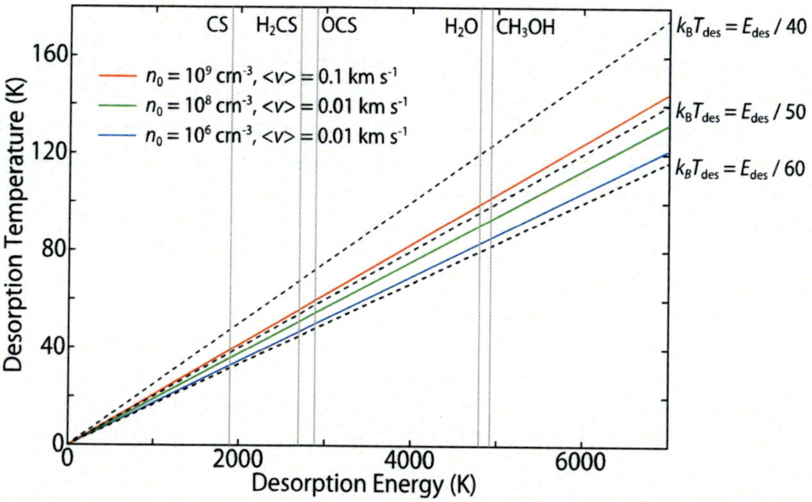

Fig. A.1 Relation between the desorption temperature and the desorption energy. Solid lines represent the plots derived from the equation (A.7), where n_0 and $< v >$ are assumed to be $(10^6 - 10^9)$ cm^{-3} and $(0.01 - 0.1)$ km s^{-1}, respectively. Dashed lines represent the plots of the following relation: $k_B T_{\rm des} = C E_{\rm des}$, where C denotes the factor of proportionality of $\frac{1}{40} - \frac{1}{60}$

Table A.1 Desorption energies and desorption temperatures of molecular species

Molecule	Desorption energy[a] $E_{\rm des} k_B^{-1}$ (K)	Desorption temperature[b] $T_{\rm des}$ (K)
H	600	11
H_2	430	8
H_2O	4800	87
CCH	2137	39
CO	1150	21
H_2CO	2050	37
CH_3OH	4930	90
H_2S	2743	50
CS	1900	35
SiO	3500	64
NH_2CHO	5556	101
H_2CS	2700	49
HCOOH	5000	91
SO	2600	47
$HCOOCH_3$	4000	73
OCS	2888	53
SO_2	5330	97

[a]Taken from UMIST database for astrochemistry ([1]; http://udfa.ajmarkwick.net/index.php)
[b]Derived from the desorption energy by using the Eq. (A.7) with the factor of 55: $k_B T_{\rm des} = \frac{E_{\rm des}}{55}$

References

1. McElroy D, Walsh C, Markwick AJ, Cordiner MA, Smith K, Millar TJ (2013) The UMIST database for astrochemistry 2012. A & A 550:A36
2. Yamamoto S (2017) Introduction to astrochemistry: chemical evolution from interstellar clouds to star and planet formation. Chapter 6

Curriculum Vitae

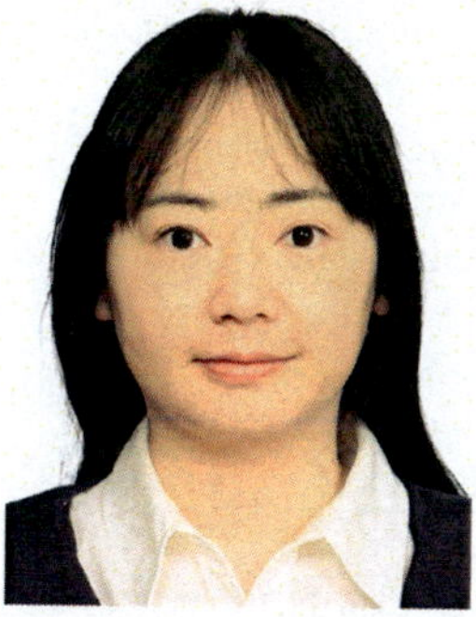

OYA, Yoko (SOMA)

Date of Birth: 18 April 1990

Citizenship: Japan

E-mail: yoko-oya@17.alumni.u-tokyo.ac.jp

Web: https://sites.google.com/view/yokooya/

Education and Degrees

2013 Mar. BSc, Department of Physics, Faculty of Science, The University of Tokyo

2015 Mar. MSc, Department of Physics, Graduate School of Science, The University of Tokyo

2017 Sep. PhD, Department of Physics, Graduate School of Science, The University of Tokyo
(Early completion)
Title of PhD Thesis: "A Few Tens au Scale Physical and Chemical Structures around Young Low-Mass Protostars"

Y. Oya, *A Few Tens au Scale Physical and Chemical Structures Around Young Low-Mass Protostars*, Springer Theses, https://doi.org/10.1007/978-981-19-1708-0

Professional Records

2015 Apr.–2017 Sep.	JSPS Research Fellowship for Young Scientists (DC1)
2017 Oct.–(present)	Assistant Professor, Department of Physics, The University of Tokyo

Major Scientific Interests

Star and Planet Formation, Interstellar Chemistry

Hornors and Awards

2010 Dec.	The 2010 ACM International Collegiate Programming Contest, Asia Kaohsiung Regional Contest, Runners-up
2010 Dec.	The 2010 ACM International Collegiate Programming Contest, Asia Tokyo Regional Contest, First Prize
2011 May	ACM International Collegiate Programming Contest 2012 World Finals, 32nd Place (Tie for 27th Place)
2014 Dec.	Poster Award, Revolution in Astronomy with ALMA -the 3rd year
2018 Mar.	The University of Tokyo President's Award for Students
2018 Mar.	The School of Science Encouragement Award (Graduate Program), The University of Tokyo
2019 Feb.	Inoue Research Award for Young Scientists

Representative Invited Talks in International Conferences

2016 Feb.	Oya, Y., et al., "A drastic change in the disk forming regions", Workshop on Astrochemistry in Star and Planet Formation, RIKEN (Japan)
2017 Jun.	Oya, Y., "L483: Warm carbon-chain chemistry source harboring Hot Corino activity", Symposium "Evolution of Molecules in Space", Hokkaido University (Japan)
2019 Jul.	Oya, Y., "Physical and Chemical Structures of Low-Mass Protostars on the Disk-Forming Scale", NAOJ Science Colloquium, NAOJ (Japan)
2019 Dec.	Oya, Y., "Sub-mm Views of Disk-Forming Regions in Young Low-Mass Protostellar Sources", EA ALMA Development Workshop, NAOJ (Japan)
2021 Jul.	Oya, Y., "Chemical Differentiation and its Relation with the Physical Structures in Disk- Forming Regions of Young Low-Mass Protostellar Sources", Astrochemical Frontiers 2021 Quarantine Edition 2, Online (keynote talk)